John Hecker

The scientific Basis of Education Demonstrated

By an Analysis of the Temperaments and of Phrenological Facts, in Connection with Mental Phenomena and the Office of the Holy Spirit in the Processes of the Mind

John Hecker

The scientific Basis of Education Demonstrated
By an Analysis of the Temperaments and of Phrenological Facts, in Connection with Mental Phenomena and the Office of the Holy Spirit in the Processes of the Mind

ISBN/EAN: 9783337280444

Printed in Europe, USA, Canada, Australia, Japan

Cover: Foto ©berggeist007 / pixelio.de

More available books at **www.hansebooks.com**

THE

SCIENTIFIC BASIS

OF

EDUCATION

DEMONSTRATED.

BY

AN ANALYSIS OF THE TEMPERAMENTS AND OF PHRENOLOGICAL
FACTS, IN CONNECTION WITH MENTAL PHENOMENA
AND THE OFFICE OF THE HOLY SPIRIT
IN THE PROCESSES OF THE MIND:

IN A SERIES OF LETTERS,

TO THE DEPARTMENT OF PUBLIC INSTRUCTION IN THE
CITY OF NEW YORK,

BY JOHN HECKER.

NEW YORK:

PUBLISHED BY THE AUTHOR, 56 RUTGERS ST.
91st YEAR OF OUR INDEPENDENCE ON THE FOURTH DAY OF JULY,
IN THE YEAR OF OUR LORD,
MDCCCLXVII.

INDEX.

A.

	PAGE
Anatomy, classification by facts of	29
Amativeness	54
Associated Apparatus	54
Aptitude,	100
Animal Magnetism,	104
Adhesiveness, how to control	130
Approbativeness	131
Action of Brain, Stomach, Liver and Lungs should be harmonious	166

B.

Boyishness of character,	41
Benevolence or Brotherly-Kindness,	91
Boys, how to treat	133
" dangers of reciprocal affection of	135
" and Girls, difference in Propensities of	136

C.

Classification,	4
" advantages of	9
" by Temperaments	11
" of Dr. Spurzheim defective	163
Children, female teachers preferable for	3
" reciprocal influence of	4
" of different temperaments to be associated	8
" importance of first impressions on	35
" " " securing affections of	70
" temperamental peculiarities of	55
" how to secure ascendancy over	69
" directions for management of	36

"   treatment of, modified by Temperaments,.................... 72
"   "   "   "   74
"   mental life of...124
"   order of development of Faculties of........................125
Combined action of the senses.. 10
Consciousness.. 20
Complexion, influence of Temperaments upon........................ 57
Child and Teacher contrasted... 71
Cerebral form, general principles of...............................81–82
Conscientiousness or Righteousness..................................... 94
Combe's Mr George translation,... 27
Cautiousness, how to control.....................................129–131
Combativeness,...130
Cluster, Intuitive and Meditative......................................101
Clusters, relative influence of different..............................143
Cluster, Conceptive how to educate the.................................145
"   "   mode of reasoning of..................................147
Clusters, proper development and culture of............................145
"   relations of..146
Causality, faculty of..141
Comparison, faculty of...141
Civilization, definition of..126
Characteristics of Propensities and Intellect.......................... 87

D.

Dalton, Dr, description of Nervous System by.......................... 19
Draper, "   "   of Nerve cells and Fibres by.................... 21
"   "   "   "   Ganglia and their relations by.................. 23
Discipline, sympathy in... 35
"   persuasive means of... 73
"   and instruction,..68–76
"   71
Destructiveness,.. 90
"   how to control..130
Development, law of...123
Diagrams, descriptions of...115
"   "   "   Profile view...118
"   "   "   Front " 119

E.

Executive force, cultivation of....................................... 36
Executiveness,.. 90
Education, order of dealing with the Faculties in..................... 34
"   present system of, defective....................................127
"   defects of special systems of...................................153
Eventuality, location of..139

INDEX.

F.

Female Teachers preferable for young children and why 3
Functions of Cerebral Ganglia 20
Faculties, distinctions between the groups of 32
    "   names given by Dr. Spurzheim 33
    "   examination of individual 36
    "   characteristic action of 37
    "   Association of 106–107
    "   practical method of demonstrating the 111
    "   boundaries of groups of, not fixed 116
    "   activity of, indicates influence 121
    "   Order of, presented by popular education 136
    "   Conceptive 140
    "   Constructive 142
    "   Combinative 143
    "   "   how to instruct the 144
    "   "   " 148
    "   Classification of, by Dr. Spurzheim 162
    "   Names and influence of Intuitive 167
Faith, ... 86
    "   Definition of 159
Form, faculty of 139

G.

Growth, laws of 2
    "   regulation of temperamental 39
Gall, Dr. mental processes of 27
    "   Nomenclature of 89
Groups and Clusters of Faculties 83
    "   relative predominance of, to be regarded 109
    "   as presented by head of Washington 115
    "   Cultivation and development of 145
Godliness, ... 90–91
General principles established by Holy Scriptures, ... 161
Girls, management of 133
    "   What Faculties are characteristic of 131

H.

Hemispheres unfolded, description of 28
    "   organic form of,   " 30
Hope, ... 95
Humility, ... 102
    " 159

I.

Interrogatories, Physiological and Phrenological.......... 13
    "   "   14
      " importance of foregoing......................... 15
Influence of organization in Spiritual Gifts,................ 26
Intercommunications of Faculties,......................... 31
Indications of Mental Character in Side View,.............. 63
      "   Front   "   64
Imitation,.. 99
Individuality,...138
Intellect, effect of overtasking the........................149
Intellectual Group and the Lawyer......................... 84

L.

Language, Phrenologically considered....................... 86
Lymphatic Temperament in Woman,........................ 53
Language, location of faculty of...........................139
    " definition of................................... 85
Lawyer and Intellectual group,............................. 84
Love,... 86

M.

Maternal sympathy,....................................... 2
Mental Sensibilities, treatment of.......................... 10
Meekness,.. 25
Motives by which children may be controlled,.............. 36
Muscular and Osseous Systems,............................ 54
Mental Capacity depends upon the brain,................... 59
Marvelousness,... 97
Mediums,...105
Mental action modified by bodily conditions,...............124
Meditative Cluster,.......................................101
Mind, laws of........... ;................................148

N.

Necessity of Spiritual Faculties to Teachers work,.......... 41
Nomenclature of Faculties, by Dr. Spurzheim,.............. 17
    "   "   " the author,........................ 17
    " importance of correct........................164

O.

Organs, relations of Faculties to... 39
" Significance of different developments of............................108
" diversity in position of... 61
" Specific location of..110
" in excess, treatment of..128

P.

Persistency,... 6
Placidity,... 6
Peculiarities of character arising from Temperaments,........................ 7
" Physical, of the organs,... 39
Principles affecting education.. 42
Physiology,... 42
Phrenology, objections to... 24
" errors and defects of... 26
" deficiencies of, in respect to temperament,........................... 67
" first established by Conceptive reasoning,............................ 47
" Gall and Spurzheim,, founders of......................................165
Phrenologic organs variable in size and position,........................... 60
" observation of character,...110
Propensities, Inheritable...127
" and Politician,... 85
" various developments of...114
" activities of, to be regarded,..116
" how to govern...132
" importance of controlling the...148
" Natural manifestations of...157
" when ruled by Intellect,..158
" should be Spiritually controlled,.....................................160
Pantomimic expression,..123
Punishments, discrimination in..134
" general observations on..150–152
Parents, duties of..135
Parentage influence of, on Intellect and Propensities,......................127
Perceptive Faculties, names of..143
" importance of...138
" " 149
" " are first educated..138
" " " 149
" how to train...137
" mode of reasoning of...147

R

Restraining Faculties	31
" "	105
Religion	43
" popular, merely intellectual	165
Religious instructions, popular errors of	154
" " value of	155
" " why irksome	40
Reflection defined	88
Reverence	91–102
Righteousness	94
Reasoning, three-fold basis of	146

S

Sluggishness, cause of	6
Sensibility, law of	9
Science of mind not purely physical	18
Sensation, definition of (Dalton)	20
Spiritual Faculties, mode of nomenclature of	17
" " importance of, to teachers	41
" " their functions	87–88
" " clusters of	101
" " the law of	103
" " names of the	90
" " composite action of the	102
" " perversions of	104
" " manifestations of	158
" " tests of predominance of	160
" " true nature and influence of	164–166
Spiritual Insight and Aptitude of teacher	98–99
" existence, evidences of	22
" Light, conditions of	25
" Life	39
" power relation of, to temperaments	48
" truths	24–25
" discernment	62
Sympathetic action and sensibility of mothers,	40
Social Science	43
Size and shape of body, affected by temperament,	57
Spiritual Insight	97
Sectarianism	104
Spiritualism	104
Spurzheim's Dr., classification of Faculties,	28
" " description of intercommunication of organs	31
" " mode of nomenclature of,	80

INDEX.

Steadfastness, Faculty of.. 93
Size indicates capacity,..120
"   location of Faculty of,...139
Secretiveness, influence of..129
"   how to control...131
Self-Esteem   " 131
Self-abasement..161
Spiritual life..156
"   Group.. 87
Soul the, is undefined.. 68
Spiritual group and Theologian.. 84

T.

Temperaments, general description of..5–9
"   Primary and combined.. 47
"   Nervous with Lithograph...48–50
"   Sanguine   "   "   51
"   Lymphatic "   "   52
"   Bilious   "   "   53
"   Combined.. 57
"   Bilious-Lymphatic... 58
"   Nervous   "   58
"   "   Bilious.. 73
"   So-called "Vital"... 55
"   Diversities of.. 63
"   present organization only... 68
"   and mental character in adults.. 80
Teacher, description of..76–78
"   temperament of.. 66
"   directions for..69–71
"   " 128
"   " 130–133
"   duties of...135
"   suggestions for..142–143
"   importance of spiritual truths to.......................................156
"   self-adaptation important to.. 69
"   gifts of, may be acquired... 77
Temperaments, development of.. 72
"   modifying influence of..122–3
"   relations of spiritual power to... 48
"   subordinate relations of.. 54
"   juvenile phases of.. 55
Temperamental observation, importance of.................................... 66
Temperaments, description of illustrations of............................... 45
Theologian and Spiritual group.. 84

V.

Volition.. 20

Veneration or Godliness.. 91
Vivacity mental, causes of.. 6

W.

Washington, Characteristics of Bust of.. 17
"   character of, mental and temperamental....................65-66
"   mental characteristics of.......................................115
"   "   "   119
"   countenance of.. 60
Wonder.. 97
Weight, location of...139

FINIS.

INTRODUCTORY NOTE

TO

THE TEACHERS.

As I lay these pages before you, the Teachers of the Public Schools of the city of New York, at the request of the Superintendent of Public Instruction, I feel that the suggestions which they contain will tend to promote the cause of Education, chiefly by promoting the material self-interest of the teacher.

So far as you are able to make practical application of the principles I discuss, they will tend to increase your ability, diminish your labor, conduce to your actual success and professional advancement, and enhance your rewards.

The peculiar power of the successful teacher consists in an ascendancy over the pupils' minds; without such ascendancy he cannot succeed, however great his learning, or however excellent his Intellectual system; with it, he can succeed, though his attainments be limited and his system crude. For every teacher knows how to acquire the knowledge he is to impart, and can conform to a prescribed system; but how intelligently to open and hold the minds of the children so that they will take it in,—this is the difficulty.

The object of the correspondence which is here laid before you, is to describe, with special reference to Education, the phenomena of human life as manifested in the physical organization, the functions of the senses, the consciousness of children, and the experience of the teacher; so that the teacher may perceive the conditions upon which this intelligent ascendancy depends.

The phenomena manifested in school children from about the age of seven years until that of about fifteen, should first be the especial study of the teacher.

During this period, the child's mental life chiefly consists of that general organic sensibility connected with the healthy or unhealthy state of the bodily functions of demand and supply, which are necessarily paramount in activity and importance at this growing period, and in the activities of the desires of the mind under impressions made through the organs of special sense,—sight, hearing, feeling, (including both the sense of touch and of muscular exertion,) taste and smell, and the reaction of the mind, in perception, through these organs.

These qualities, which may be termed the *sensuous* susceptibilities of the mind, engross nearly all its activities during this period of growth; and there is no other avenue by which the teacher can successfully approach and control the children's minds.

The accuracy, vividness, and strength of the mental perceptions which he arouses in their minds, will depend upon the vigor of the senses, and the skill with which he deals with them; while the truthfulness, clearness, and permanence of the resulting benefits to be evolved in the higher processes of education will essentially depend in part upon the quality of these sensuous rudimental perceptions.

The second stage of development next demands the teacher's attention. As the characteristics of childhood merge in the desires of youth, the higher mental sensibilities and intellectual powers, which at first had only an incidental and secondary character, become more prominent by the long operation of the laws of bodily growth and mental development. The temperamental disposition and the mental character now begin to assume a settled form, and the teacher, though he must not disregard the desires of the sensuous conditions, is no longer dependent chiefly on them, but must obtain and wield his influence through temperamental and mental laws.

But besides these changes in the course of development, every teacher observes the endless diversity of disposition and capacity in pupils of the same age,—diversities which constantly confuse his discipline and instruction.

Some children are more sensitive than others to their physical condition and to external objects. This is the first point which the teacher should observe in examining a scholar, for it indicates the acuteness of the organs of special sense and of the general sensibility of the desires; although the quality of these sensibilities varies according to the temperament.

Again, some children are mentally disposed, the nervous forces tending to activity in the cerebral hemispheres, and these are apt, or quick-minded; others are bodily disposed, the nervous forces being more engrossed in the lower centres, giving a more gross or physical nature, and these are slow to learn. Both these qualities are manifested in the action of the senses. Some are unimpressible, but when impressed, retentive; others volatile and incessantly diverted. These and similar general characteristics, which often become more and more developed with growth, depend on temperament.

Again, some children have strong desires and propensities manifested in social gregariousness, in selfish interests, or in segregation and evil indulgences; some being high-spirited and proud, others sensitive and vain, others sensual and vicious.

Other children are characterized more by intellectual qualities than by feelings and desires, some being intelligent observers without much power of analysis in reasoning, some readily understanding the philosophy of a subject but not easily acquiring the facts, and some with less knowledge, are more ingenious in giving structural form to their ideas. And, lastly, a small class of minds in childhood are characterized by moral rather than by intellectual or passional qualities. All these and other mental diversities are connected with the functions of the brain.

All these differences of disposition and character, both Mental and Temperamental, are indicated by Physiological and Phrenological peculiarities. They can be definitely analyzed and described; and the teacher who will give the necessary attention to these external manifestations, and the subjective qualities indicated thereby, may learn to understand his own peculiar powers, and intelligently to observe the differences in his scholars, so as to adapt himself to them, and treat each as his nature requires.

Some persons are gifted with such a degree of Physiologic and mental sensibility in sympathy with children, that they are by nature successful teachers, they know not why. But those who are not thus gifted may acquire much of the same power by attention to the conditions on which it depends.

He who undertakes to teach children should first enquire whether he is gifted to do so by his own powers of special sense and his general organization. If his senses, especially sight and hearing, are not quick and acute, and his mental sensibilities are not in sympathy with the peculiar sensuousness of childhood, he should either correct his deficiency or apply himself to the instruction of more advanced scholars.

Efficient teaching of children in Primary schools depends less on cultivated intellect and special knowledge than on the teacher's successful adaptation of himself to the sensuous nature of childhood. This is a fundamental truth, which should be seriously considered by the Superintendents of our Public schools, in the selection of teachers, and in training in the Normal School.

The Primary school teacher should prepare for the work of the schoolroom by studying and rehearsing his external methods, in signals, demeanor, bearing, gesticulation, tone, address, and all visible and audible demonstrations, that he may keep the senses of the children on the alert, and when he speaks, may touch the whole, like an electric battery.

The senses act simultaneously, and may thus be said to reciprocate with each other through the desires of the passions, they being the common centre of the Physiologic forces of the brain. From the reciprocal influence of the senses, results the importance of exercising both sight and hearing together, as in music and in teaching with the blackboard: and the same principle should be extended to the other senses as is already practised in what is called Object-teaching. While dealing with any one or more senses, the teacher should take care that all the other sensibilities are under conditions favorable to the impressions to be made, and therefore he must be as much alive in all the physical sensibilities as the children are; must have regard to their food, exercise, and repose, and, during the period of instruction, be constantly attentive to light, air, temperature, their bodily attitudes, and every physical want.

Having by these means secured the sensuous and perceptive attention of the children, he must use this power faithfully and efficiently to awaken mental life; he will fail of usefulness unless he wisely directs to mental functions the sensuous energies with which he is thus dealing.

In order to secure affection, and call to his aid the peculiar disposition of each, he must understand their temperamental characteristics, that is, the phase of the bodily or vegetative functions by which the brain is supported and sustained. In order to develop harmoniously the mental powers, to apply the instruction appropriate for minds of different cast, and qualify each for success in life, according to his or her capacity, and in order to understand the motives with which he has to deal, he must study the mental qualities manifested in physical organization. By this knowledge alone can he attain the conscious ability to modulate his disposition and adapt his instructions as his task requires.

Throughout the whole course he must intelligently enter into the feelings of desire which animate children and youth; and must suffer with them, and if possible suffer for them, that he may keep their sympathy.

Love for a good mother lingers in a depraved mind long after other moral qualities have been abandoned; this results from what the mother has suffered with and for her children. If the teacher awakens the sensibilities of scholar's minds by some self-denying suffering for them, he secures the most important condition of control, which is sensuous sympathy through the Perceptive faculties and Propensities; and by this means may begin to evoke a higher sympathy in those moral and Spiritual qualities which in children are immature, and on which, by reason of the sensuously engrossed state of the natural mind, the teacher cannot rely.

If punishment is to be applied, it should be administered sympathetically. The mother's heart pleads for the boy, whatever he has done, and the teacher's heart should do the same. He must always be conscious that it is not in human nature to punish with absolute justice, and must, therefore deal in charity, and make his kindness manifest in the act of discipline, ever remembering that the child has a soul as well as intelligence to care for.

The Physiologic and Phrenologic principles underlying the distinctions of character, and the sensuous and mental sympathies, I have briefly described, afford to the teacher the means of possessing the minds of his scholars in lively unison with his own. In this state of *rapport*, he may exercise their minds with whatever affects his own, and may impart instruction with pleasure to himself and to them. A conscious realization of this knowledge and feeling will be a source of the greatest delight in his vocation.

In the practical administration of discipline and instruction, we should never forget those truths originating in the Spiritual nature of man, which an inquiry into his organization and his history demonstrates, in which are found the ultimate object of all education, and in which alone the teacher can find the power necessary for his repose.

TO THE TEACHERS.

As these letters have originated in the form of questions, I will add, that if in anything I have not been sufficiently explicit for your practical purposes, or if doubts arise, or branches of the subject I may not have touched upon require elucidation, I shall cheerfully respond to any additional inquiries you may be pleased to put to me.

I am aware that Phrenology will be regarded by many of my readers, as it is by many scientific men, as unsupported by Physiology. The position which Phrenology, as a system of mental science, occupies in the estimation of the leading scientific men of our day cannot be better stated than in the words of Sir Henry Maudsley. (Physiology and Pathology of the Mind. Lond., 1867).

"It is extremely probable," says he, "that different convolutions of the brain do subserve different functions in our mental life; but the precise mapping out of the cerebral surface and the classification of the mental faculties which the phrenologists have rashly made, will not bear scientific examination. That the broad and prominent forehead indicates great intellectual powers was believed in Greece, and is commonly accepted as true now. * * * There is some reason to believe also, that the upper part of the brain and the posterior lobes have more to do with feeling than with the understanding. * * * On the whole, it must be confessed that, so far, we have not any certain and definite knowledge of the functions of the different parts of the cerebral convolutions.

This is about as far as scientific Physiologists are yet agreed.

The popular estimation of Phrenology accepts it as a common-sense analysis of mental faculties and nomenclature, which in the words of Archbishop Whately, is, " far more logical, accurate and convenient than those of Locke, Stewart, and other writers of their school."

It is admitted, however, that Phrenology has performed the service of pointing out and inaugurating the true method of investigating the mind; and the researches of Bain, Spencer, Laycock, Maudsley, and Carpenter, in England, and of Draper and Dalton in our own country, go to confirm the two fundamental principles upon which Dr. Gall's system was based, viz: that different faculties or modes of mental phenomena are localized in different parts of the cerebral hemispheres, and that power of mind or faculty, is, other things being equal, dependent on or indicated by size or development of the brain or part of the brain through which it is manifested.

Without pursuing this topic further here, I would refer you to the Encyclopedia Britannica, which in its last edition has cancelled the decisive condemnation it formerly pronounced upon Phrenology, and says that its claims are now *sub judice*. The account which it gives of the principles of Phrenology is one of the best statements of the general doctrine which I have seen.

In these letters, I shall explain why Drs. Gall and Spurzheim failed to establish the science of the mind beyond controversy; and I shall endeavor to present the subject of cerebral form in such a definite, yet natural and practical manner, that a careful reader can verify, in his own observations, the general truth of these principles, and I shall apply them especially to Education, for the benefit of the Teacher.

The progress of Physiological science, thus applied, renders it possible to delineate the organic phases under which all diversities of disposition and capacity are manifested. The science of the mind, when thus understood, will define organically every peculiar phase of thought, and thus solve every Philosophic controversy, and will elucidate, at once, the diversities of opinion which are manifested by diversities of organization, and the unity of the Truth, which is only manifested through the Spiritual nature of man by the grace of Almighty God.

The future welfare of Society is dependent upon the fidelity and success of the education which you are administering. In a vast and cosmopolitan population, the Public Instruction of New York maintains a character of true universality, receiving into its care all comers of whatever class, creed, nation, sect, race, or color, and dispenses its beneficent influence throughout the community, free as the air we breathe.

It provides edifices, with halls and class-rooms, for the accommodation of two hundred thousand pupils, yearly, and this vast number, gathered from every grade of society, presenting every diversity of capacity and disposition, are brought together in large masses of from one hundred to six hundred, under perfect discipline, and swayed as one, by the voice or signal of a single teacher. You are supplied with intellectual resources and material facilities for your work, and your labors are ably directed and superintended. The institutions of Public Instruction of this city are looked to from other States and countries as the worthy Metropolitan representatives of the American Educational System.

As science advances in the discovery of Truth, Education, which is the popular absorption and assimilation of Truth, must also advance.

To you, then, whose noble and honored function it is to lay hold upon the truths which science develops, and to bring them to their utilization by infusing them into the common knowledge of the community, I appeal to investigate these principles, to put them to the crucial test of practice in teaching, and join with me in advancing still further the cause of Education by bringing to the work a scientific knowledge of the human mind. J. H.

56 Rutgers St.—NEW YORK, *June,* 1867.

LETTERS

FROM LEADING SCHOOL SUPERINTENDENTS.

The following extracts are taken from letters received from prominent literary gentlemen and Educators, with whom the Author has been in correspondence during the progress of this work.

DEPARTMENT OF PUBLIC INSTRUCTION,
SUPERINTENDENT'S OFFICE, 146 GRAND ST.,
New York, *Sept.* 26, 1866.

JOHN HECKER, Esq.
INSPECTOR, 2ND DISTRICT.

DEAR SIR :—In accordance with your request, I have carefully perused the full exposition of your views on the subject of classification of the pupils of our Public Schools with reference to the respective temperaments as contained in your answers to the inquiries propounded by Assistant Superintendent Kiddle. Those views are in my judgment of the highest practical importance as well to teachers as pupils—based upon the soundest principles of physical, mental and moral science, and admitting of practical application by every teacher who will take the pains of acquainting himself or herself with the principles upon which they are founded and from whence they are legitimately and clearly deduced. I should be glad to see them placed in the hands of all our teachers, in the assured conviction that their general adoption would essentially advance the best interests of education; while at the same time the full and minute analysis given by you of the philosophy and science of instruction, of the grounds upon which it rests, the responsibilities it involves, the duties it demands, and the constant reference to the immortal nature of the beings upon which it acts cannot fail of commending itself to the profoundest attention and regard of every faithful and conscientious teacher.

Very respectfully and truly,
Your friend,
S. S. RANDALL,
CITY SUPERINTENDENT.

NEW YORK, May 21, 1867.

JOHN HECKER, Esq.

MY DEAR SIR:—I have deferred until the present time formally expressing to you my opinion with regard to the views set forth in your reply to the interrogatories proposed by me some time ago, because I was desirous of having before me a complete statement of the theory and plan upon which in your judgment educational processes should be based.

This, I believe, has now been done. The replies which have been submitted are full and comprehensive, containing an enumeration and elucidation of the principles which should guide the teacher in the selection of general methods of operation, and in making those discriminations which are required to enable him to adapt these methods to the peculiarities of individual character.

The need of such an adaptation of general methods and processes must always have proved a source of perplexity and anxiety to the intelligent and conscientious teacher, convinced that the true measure of his success must be the extent of his accuracy in making the discriminations required for a wise selection of agencies both in discipline and instruction. The difficulty, however, has been, that trusting to the imperfect criterion of character which a mere observation of conduct affords, the best teachers must have very frequently and sadly erred, and consequently have fallen short of that full and true success which can be attained only by a surer and deeper insight into the individual characteristics of those who are to be taught.

The more reliable guide to such an insight afforded by the correct observation of temperamental and cerebral peculiarities, for which clear and explicit directions are given in your letters, must, without doubt, if it be faithfully applied, greatly facilitate and increase the success of the wisest and most experienced educators who have hitherto worked without it.

It was this consideration that first attracted my attention to your views. Engaged in the work of education from a very early period of my life, so circumstanced as to be obliged to take notice of the various methods employed by teachers, and often painfully impressed with their inefficiency and want of adaptation, sometimes to the general objects of education, much oftener to the specialties of individual organization, I at once saw that, in the principles of a rational physiological and phrenological system, clearly understood and skilfully applied, was to be found the instrument of a more thorough and more assured success.

Through your means, this instrument will now be placed in the hands of every teacher whom this published correspondence shall reach; and a solemn obligation will rest upon him to study the principles expounded in these letters, the directions there given for their proper application, and the philosophy underlying them, which is there discussed and explained;

and I trust the time is not far distant when it shall be deemed of the highest importance, as a part of the professional preparation of the teacher, that he should acquire not only a thorough knowledge of these truths, but practical skill in applying them, so that he may be able readily to discern the diverse tendencies and capabilities of the young minds committed to his guidance and training.

I feel also that you have done an especial service to the cause of education, as well as to that of religion and sound morality by inculcating it as an indispensable requirement, that those who are to train the youth of a Christian community, should cultivate their higher nature—those capacities for spiritual enlightenment, by which, in accordance with that pure Christian philosophy which you have so well interwoven with the theory of our mental and physical organization, the human character is exalted to higher views, holier aspirations, and a nobler self-devotion to duty.

In thus expressing my cordial assent to the educational philosophy embodied in these letters, and my appreciation of the practical value of the truths which it comprehends, and the precepts which it dictates, permit me also to express my sense of the earnest philanthropy and public spirit which your exertions so long continued in this work have demonstrated. That those for whom it is intended may accept it cheerfully, and employ it faithfully and wisely, is the sincere wish of

Yours truly and Respectfully.

HENRY KIDDLE.

From Professor A. Parish,

SUPERINTEDENT OF SCHOOLS,
NEW HAVEN, CONN.

"I must frankly say to you that I have never had much faith in Phrenology, as it has been promulgated during the past quarter of a century; indeed my prejudice has been strongly adverse to it. When I perused your paper, and re-read parts of it again and again, as I have with increasing interest, my conclusion was that you have struck a new vein on the subject—rather have presented the subject in an aspect so new and given a character of *practicality* to it, that I am quite desirous of knowing what use can be made of it.

"Human nature must be read through the *actions* and *words* of the individual; and these are prompted and governed in no small degree by the character of the *physical* man. Hence the study of Temperaments becomes important. If I know the temperament of a child, I know how to approach him to accomplish a given object, to what motives to appeal, what influences to bring to bear upon him, etc. In this view of the matter I am interested to know more of your views and the whole subject. If it can be brought to the comprehension of the ordinary teacher and can be reduced to a practical application in the school-room, it is obvious that the matter of school discipline must be greatly modified if not revolutionized, and all the processes of instruction materially facilitated.

March 14th, 1867.

From G. B. Sears, Esq.,

SUPERINTENDENT PUBLIC SCHOOLS,
NEWARK, N. J.

I have perused your "Letters" with much interest and hope a copy may find a place in the Library of every person who has any thing to do with the instruction and discipline of children. That there is a "Scientific Basis" upon which a system of instruction rests—that God has not committed so great a work as the development of immortal natures to human hands without some discernible law of action I have long been convinced. These "Letters" appear to me to be a step in the right direction. They reveal the legitimate avenues by which the teacher may obtain access to the source of moral and mental action and educate the various faculties in harmony with each peculiar constitution—doing violence to none.

"The 'Letters' suggest a wide field and a very interesting one which our teachers should carefully explore. Teachers should apply these principles in the school-room, where they may be tested upon every variety of temperament and condition, but only after a thorough understanding of them themselves. These 'Letters' strike at the very foundation of self-knowledge and speak to each one emphatically, 'Know thyself.' When teachers shall take the pains to acquaint themselves with their own temperaments and peculiar constitutions, and apply also the principles here laid down to acquire a knowledge of children under their charge, the main obstacles in the way of a symmetrical and beautiful development of human character will have been removed."

February 15th, 1867.

From C. R. Abbot, Esq.,

SUPERINTENDENT OF SCHOOLS,
KINGSTON. N. Y.

"After reading the 'Pamphlet' I intended to submit some thoughts on the subject. On some of the points taken we should disagree. These, however, may be unimportant. I am heartily glad you have so written and am grateful for the copy sent. I have a profound interest in this subject. To my mind it affords the means to overcome many, very many of the difficulties with which teachers and supervisors of educational institutions now vainly *contend*. For many years I have recognized the truths stated by you, but I am indebted to you for putting *words* into my mouth. Too many of our teachers are blind leaders. They have no just conception of a perfect human character, and when they take charge of children whose minds and affections are stunted and deformed, they only make the matter worse by their unskilful management. With your permission I will keep the 'Pamphlet' a time longer, and shall be thankful for more light on the subject."

March 16*th*, 1867.

From the Rev. Edward Ballard, D. D.,

STATE SUPERINTENDENT OF SCHOOLS,
BRUNSWICK, MAINE.

* * "As to your proposals relative to an improvement in the classification of scholars by temperament: I can say that as presented by your 'Pamphlet,' it has much to recommend it; and where the judicious teacher can make the due selection and arrangement, his work will be lightened, the tuition be the more easily applied, the pupils the more interested and the less fretted under the restraints of the school-room and class exercises, and success the more certain to be attained. Temperamental classification is worthy of a fair trial.

"In regard to the Phrenological part of the system, I am pleased that you are placing the science where it can be made to have a religious bearing; and where too, in the hands of the teacher acquainted with its principles it can be made to regulate the tuition to be poured in and the development to be brought out." * * *

March 18*th*, 1867.

From the Hon. John A. Norris,

SUPERINTENDENT PUBLIC INSTRUCTION,
COLUMBUS, OHIO.

* * "Those who are engaged in the work of advancing the interests of Common Schools cannot but be forcibly impressed with the soundness of the principles upon which your discussions are based and the radical changes in the prevailing systems of classification in our Public Schools that must inevitably follow, if the conclusions set forth by you are successfully established. I am not now prepared to discuss your views nor even to make suggestions in regard to them, but their importance is such that I hope to be permitted to retain the papers you have already so kindly furnished me. May I not hope also that your discussions, accompanied by suitable illustrations of the temperaments, will be given to the public at no distant day?

March 27th, 1867.

From the Hon. G. W. Hoss,

SUPERINTENDENT PUBLIC INSTRUCTION,
INDIANAPOLIS, IND.

* * "Your positions are in many particulars *new*, at least to me; hence the tendency is to accept *slowly*. Yet notwithstanding this fact, I am prepared to announce my convictions, namely, that your doctrine of Temperaments is at once sound and philosophic; hence should be recognized in all philosophic systems of education. I thank you in the name of an humble believer in Christianity, that you declare the importance of *moral culture*; that you clearly recognize the operations of the Holy Spirit. In brief and fine, I can condense my statement into a single line, by saying that I am of the opinion that the doctrine set forth in these pages is to be *The new Evangel of Education*."

March 29th, 1867.

SCHOOL SUPERINTENDENTS.

From the Hon. Daniel Stevenson.

SUPERINTENDENT PUBLIC INSTRUCTION.
FRANKFORT, KY.

"My very limited knowledge of the science of Phrenology would render it improper for me to express an opinion in regard to some of the views presented by you. I may say, however, that I have read the "Pamphlet" with very great interest, and, I will add, with profit. With its general spirit and design I was much pleased. The discussion cannot fail to be of advantage to the educational interests of the country."
March 26th, 1867.

From the Rt. Rev. J. H. Hopkins, DD. LLD.

BISHOP OF VERMONT,
BURLINGTON, VT.

"As a whole I regard your work as a very admirable contribution to the philosophy of true education, which presents a more profound and scientific view of the subject than any other work within my knowledge, and must as it seems to me produce a highly beneficial influence on the minds of the thoughtful."
March 30th, 1867.

From John D. Philbrick, Esq.,

SUPERINTENDENT OF SCHOOLS,
BOSTON, MASS.

"I have not yet had time enough to read your 'Pamphlet' with sufficient care to warrant a decisive opinion on its doctrines as a whole, but I send this preliminary note to tell you I am extremely interested in your inquiries, and that I am very favorably impressed with the spirit and tone of what you have presented. I am already an admirer of Spurzheim, and it seems to me that, in some respects at least, you have improved on him."
March 12th, 1867.

From *Albert G. Boyden, Esq.*,

PRINCIPAL STATE NORMAL SCHOOL,
BRIDGEWATER, MASS.

* * "The prominent place which you have given to the Spiritual part of our nature in your discussion of the subject places it upon the right basis, in my estimation. It seems to me that your views will be of great service to every intelligent teacher. Nothing pleases me more than to find any thing which will facilitate the right education of the young, and especially that which conduces to the higher development of our Spiritual Nature."

March 16th, 1867.

From the *Hon. William R. White,*

SUPERINTENDENT PUBLIC INSTRUCTION,
WHEELING, WEST VA.

* * "In adapting myself to the mental peculiarities of my pupils, I have, (intuitively, rather than from any scientific acquaintance with Phrenological developments), acted in accordance with many of the suggestions which I find in your 'Letters.'

"That the true teacher should be acquainted with the laws of growth governing the body and mind and soul is rendered so apparent in your remarks, that I cannot but wish that your work may find its way into every school."

March 15th, 1867.

CONTENTS.

LETTER I.

REQUEST FOR SUGGESTIONS.

LETTER II.

TEMPERAMENTAL CLASSIFICATION OF PUPILS.

Importance of Laws of Growth as connected with Education—Maternal Sympathy—Female teachers preferable for young children; and Why?—The Spoiled Child—Reciprocal influence of Children in large families—Classification—Leading Temperaments—Physiological Origin—Causes of Mental Vivacity, Persistency, Placidity, Sluggishness—Peculiarities of character, Mental and Physical, arising from Temperaments—Advantages of Mingling Children of different Temperaments—Advantages of Classification—Law of Sensibility—Treatment of the Mental Sensibilities—Combined action of the Senses—Suggestions for Classification by Temperaments..p. 12

LETTER III.

INTERROGATORIES, PHYSIOLOGICAL AND PHRENOLOGICAL.

LETTER IV.

PHYSICAL AND SPIRITUAL LAWS OF MIND.

Importance of Foregoing Interrogatories—Phrenologic Bust of Washington, its Physiognomic Characteristics—Characteristics of this Bust of Washington compared with other Phrenologic Busts—Science of Mind not Purely Physical—Dr. Dalton's Physiological Description of the Functions of the Nervous System—Functions of Cerebral Ganglia, Sensation, Consciousness, Volition—Dr. Draper's Description of Nerve Cells and Nerve Fibres—Evidences of Spiritual Existence—Dr. Draper's Description of the Sensory and Motor Ganglia and their relation to the Hemispheres—Mr. Combe's Description—Objections to Phrenology—Conditions of Spiritual Light—Meekness—Influence of Organization on Spiritual Gifts—Deficiencies of Phrenologists—Character of Dr. Gall's Mental Processes—Character of Dr. Spurzheim's Classification—Description of the Hemisphere unfolded—Classification by the Facts of Anatomy and of Human Life—Description of the Organic Form of the Hemispheres—The Faculties of Restraint—Intercommunications of the Faculties—Fundamental Distinction between the three Groups of Faculties—Classification of the Faculties according to the foregoing Views—Order of dealing with the Faculties in Education—The Restraining Faculties—First impressions upon the Scholar by the Teacher—Sympathy in Discipline—Motives by which Children may be controlled—Cultivation of Executive Force—Faculties may be considered Individually; but their action is Associate—Physical peculiarities of the Organs—Examination of Individual Faculties—Regulation of Temperamental Growth—Relation of Faculties and Organs—Spiritual Life—Sympathetic Action—The Mother's Sensibility—Why Religious influence is Irksome—Necessity of Spiritual Faculties to the Teacher's work—Boyishness of Character—Principles affecting the scheme of Education—The domain of Physiology—The Domain and Processes of Instruction—Social Science—Religion—The Domain of Religion and the Church..............p. 44

LETTER V.

THE TEMPERAMENTS.

Description of the Colored Illustrations of the Temperaments—What is "Temperament"—Primary and Combined Temperaments—What is meant by Combined Temperament—Relation of Spiritual Power to the Temperaments—The Nervous Temperament—Organic condition of the Nervous Temperament—External Indications of the Sanguine Temperament—External Indications of the Lymphatic Temperament—Lymphatic Temperament in Woman—External Indications of the Bilious Temperament—Auxiliary Apparatus—Muscular and Osseous Systems—Amativeness—The so-called "Vital Temperament"—Temperamental Peculiarities in Childhood—Manner in which the Temperaments may be described in the combined form—Complexion—Size and Shape of body—The Nervous-Bilious Temperament—The Nervous-Lymphatic Temperament—The Bilious-Lymphatic—What is the most favorable Temperament—Mental Capacity depends on the Brain—Washington's Countenance—Phrenologic organs are variable in size and position—Diversity in the Associate position of the Organs—The mind not to be arbitrarily measured—Spiritual Discernment—Modifying influence of Temperament—Indications of Mental Character in the side view—Diversities of Temperament—Indications of Mental Character in the front view—Diversities of Temperament—The Character of George Washington described, Temperamentally and Mentally—Successive changes in Washington's character—The Teacher's Temperament—Deficiencies in Phrenologic systems in respect to Temperament—The Temperaments present the Organization, only; the soul is undefined—How the Teacher may secure Ascendancy over his class—Self-adaptation—What Children give attention readily—What are persistent and retentive—Securing affection—How to address Children—Contrast between the Child and the Teacher—Discipline—Different children susceptible to different motives—The development of the Temperaments—Persuasive means of Discipline—Nervous-Bilious Temperament—Suggestions as to Modifications in Treatment—The sports of children—The choice of a vocation—Temperamental adaptation to various callings—The description of the Teacher—The gifts of the successful teacher may be acquired—Teaching an exalted Profesion—Should command best talent and largest compensation.........p. 78

LETTER VI.

THE MIND.

Recapitulation of previous letters—The temperaments and Mental Character delineated as seen in the Adult—General Principles of Cerebral Form—The Phrenologic Bust of Washington—Groups and Clusters of Faculties—The Intellectual Group and the Lawyer—The Spiritual Group and the Theologian—The Propensities and the Politician—The use of Language, Phrenologically considered—Love, Faith, Patience, Joy—The Propensities and the Intellect—Their general characteristics—The Spiritual Group—Reflection defined—Manifestation of the Truth by the Spiritual Faculties—The Principles of Phrenologic Nomenclature—Destructiveness—Executiveness—Names of Spiritual Faculties Godliness, Reverence—Veneration or Godliness—Benevolence or Brotherly-Kindness—The Faculty of Brotherly-Kindness—The Faculty of Steadfastness—Conscientiousness or Righteousness—The Faculty of Righteousness—Hope or Hopefulness—The Faculty of Hopefulness—Wonder, Marvelousness, or Spiritual Insight—The Faculty of Spiritual Insight—Imitation or Aptitude—The Faculty of Aptitude—Spiritual Insight and Aptitude of the Teacher—The Clusters of the Spiritual Faculties; Intuitive and Meditative—The Knowledge of God—Humility and Reverence—The Special gifts of the Holy Spirit—The law of the Spiritual Faculties—Sectarianism, Animal Magnetism, Spiritualism—"Mediums"—The Restraining Faculties and the Law of Association—Mode in which the Organs combine in their development—Mutual influence of Associated Faculties—Significance of Organs whether prominent or laterally expanded—The relative predominance of groups must be regarded—How to observe character Phrenologically—General form of Cerebral instrument—Indications resulting from various Cerebral forms—Influence of Temperaments—Various developments of the Propensities—General outline of the head first to be ascertained—Description of Diagrams—Mental Characteristics of Washington—Boundaries of Groups not fixed—Activities of the Propensities must be regarded—Indications resulting from predominance of various Groups and Organs—Description of Profile view—Description of Front view—Front view continued—Character of Washington as indicated by Cerebral form—Size indicates capacity—Temperamental indications—Harmony of Cerebral parts—Degree of Fibrous development—Activity of the Faculties is the measure of influence—Necessity of awakening Spiritual Faculties—Modifying effects of Temperaments—Perceptive influence of different Temperaments—Pantomimic expression—Law of due development—Bodily conditions modify mental action—The mental life of the child—General law of the development of the Faculties in children—Diversities of this law—Development of Faculties dependent on Climate, Civilization, Inheritance—Influence of Public Schools—Intellectual Faculties and Propensities inheritable—Present methods of Education defective—Directions for the Teacher—Influence of Cautiousness and Secretiveness, when developed in a greater or less degree—Instructions for Teacher—When Destructiveness, Combativeness or Adhesiveness are in excess?—How to manage excessive Secretiveness, Approbativeness, or Self-Esteem—Directions when Propensities are large and active, or small and sluggish—What the Teacher must ascertain—Difference of treatment for Boys and Girls—Importance of securing attention—Necessity of intelligent discrimination in Punishments—Dangers of reciprocal affection among boys—Duty of Parents and Teachers—Order of Faculties as presented by popular Education—Difference between Boys and Girls—" Girl-boy" and " Masculine-girl"—Perceptive Faculties must be studied by Teacher—Individuality—Importance and advantage of this Faculty to Teacher—Form and Size—Their position—Teaching by Black-board and Sound successful—How to train the Perceptive Faculties—Conceptive Faculties—Comparison and Causality, the Faculties of Reason—Their relations with other clusters—Suggestions for Teachers—Proper method of training Faculties—Reciprocal influence of the different clusters—Sports of Children—Hints for instruction when different clusters predominate—How to cultivate the Intellectual Group—Proper order of their development—The Bases of reasoning are three-fold—When reasoning commences—Relations of the Clusters—Perceptive Faculties first developed—How Perceptive and how Conceptive Faculties reason—Laws of the Mind—Importance of controlling the Propensities—Effect of overtasking the Intellect—Importance of the Perceptive Faculties—Suggestions as to methods of Punishment—How to treat Indolence—Organization and Temperament to be regarded—How to overcome Disobedience—Insubordination—Value of a generous *Esprit de Corps*—Why special systems of Education are unsuccessful—Suggestions for moral Training—Why Religion is generally distasteful to young men—Value of Religious Teaching—Moral Faculties trained by spiritually-minded Teachers only—Importance of Spiritual truths to the Teacher—The Beatitudes—Natural manifestations of the Propensities—Manifestations of Spiritual Faculties; of the Intellect when ruled by the Propensities—Faith, its definition—Humility—The rich Young Ruler—Tests of the predominance of Spiritual Faculties—Propensities to be under Spiritual guidance—Self-abasement—Principles established by Holy Scriptures—Dr. Spurzheim's Classification—Defects of Classification—Error as to nature of Spiritual Faculties—" Blind Sentiments"—Necessity of correct Nomenclature—True nature of Spiritual Faculties must be recognized—Popular Religion merely Intellectual—Drs. Gall and Spurzheim Founders of Phrenology—Harmonious action of Brain, Stomach, Lungs and Liver important—Spiritual Faculties—Influence of Intuitive Faculties—Necessity for more thorough knowledge of the mind.......................... p. 167

APPENDIX.

The Will—Consciousness—Social Organization—The Church—Conversion—The Standard of Truth.

THE SCIENTIFIC BASIS

OF

EDUCATION.

[*From Mr. Kiddle to Mr. Hecker.*]

DEPARTMENT OF PUBLIC INSTRUCTION,
SUPERINTENDENT'S OFFICE, 146 GRAND ST.,
May 24, 1865.

JOHN HECKER, Esq.,

MY DEAR SIR:

I have listened with much interest to the views which you have, from time to time, presented as to the need of modifying the processes of discipline and instruction employed in the Public Schools of the city. Very much of the criticism which you have made as the result of your observations, I deem so just and important as to demand an immediate attention, with the view to correction and improvement.

I would, therefore, request you to state, in writing, the particular methods of reform or modification which appear to you feasible under the circumstances?

Such an exposition, I feel, will be an important aid in suggesting and carrying out such measures as will prove of benefit to the schools. Trusting you will be able to comply with this request at an early day, I am

Yours truly,

HENRY KIDDLE

[*From Mr. Hecker to Mr. Kiddle.*]

NEW YORK, *June*, 1865.

HENRY KIDDLE, Esq.,

DEAR SIR:—The opportunity which your late note of inquiry affords me of laying before you, in writing, some views connected with the subject of practical education in our Ward Schools, I very gladly accept. In the discharge of the duties devolved upon me, as Public School Inspector of the Second District, it is my desire to co-operate in every way in my power, with my co-laborers in the department; and I feel it to be my immediate duty to lay before you such suggestions as an experience of thirty-five years, spent in closely studying the nature of man, enables me to make.

In the first place, the most important fact which has been apparent to me, after having examined the eighteen Public Schools of my district, and acquired some general knowledge of the magnitude and importance of the system of public instruction throughout the whole city, (at the head of which Mr. Randall and yourself, with Messrs. Calkins, Jones, and Seton, are practically placed) is, that the *laws of mental development*, and the *laws of growth* which affect that development, seem unobserved both in the general organization of the system, and by the teachers in practically administering it. There is scarcely any recognition of the existence of these laws in the nature of the child.

Education is not alone concerned with the imparting of information. In order that the information may be received, there must be a certain growth or development; and to guide and assist the divinely ordained process of growth both of body and mind, is a part of the duty of the educator; because this right development is required, both for the successful communication of knowledge, and for the ultimate welfare of the subject, without which the acquisition of knowledge may prove a curse instead of a blessing.

Therefore, the methods of education should be based upon an intelligent recognition of the laws of growth, and a sympathy with them, and an adaptation to them.

Growth, it is true, is considered by teachers, to a certain extent; but it is measured, usually, by size, or rather height of body, and by the existing amount of information possessed by the child. These are important circumstances to be taken into account, in estimating the growth of a child; but they are not to be relied upon as guides. There exist, in the

Maternal Sympathy. Female Teachers preferable for Young Children;—and Why?

nature of the child, certain laws of growth, and these, so far as they relate to the bodily condition and development, depend on what are denominated the temperaments. As the brain is dependent on the body, which is to it, what the soil is to the plant, these temperamental laws are of fundamental importance in dealing with the growth of the mind; and a sensuous understanding of them is essential to give the teacher that untiring sympathy with the children which ought to be possessed.

The earliest phenomenal appearance of organic mental life is at birth, in the infant's crying. This is caused by the inhaling and exhaling of the atmosphere, expanding the cells of the lungs, producing pain. Here the first attention of the mother is called to her offspring in sympathy. I here use the word "sympathy" in a more definite and literal sense than the ordinary acceptation of the term, to denote an interpenetrating, intelligent, mutual sensibility peculiar to a mother, and which has its origin in the sufferings belonging to maternity, though those sufferings are suspended by the love she bears her child in the predetermined organic disposition of the social part of her mental nature.

The mother of the child is the ordained means by which the child is properly to be cared for, up to the age of seven years. Throughout this early period, a good mother will always use, as the leading means of her care, the same sympathy, together with, however, the necessary punishments to awaken a proper degree of fear and respect. The teacher ought to take up the work at the point where the mother leaves it, and should deal with the same means of obtaining submission:—first, sympathy, then, also, fear and respect. It is because of the necessity for this sympathy, that experience and practical knowledge in our primary schools place children of a tender age in charge of female teachers. The female is peculiarly gifted in her sensuous organization. It is owing to her characteristic temperamental quality, that she is so peculiarly sensitive to physical as well as mental impressions. Her daily intercourse with children, and her intuitive sympathy in their wants and necessities, constantly strengthen this predisposition of her sensuous and sympathetic nature. Hence too, she arrives promptly at conclusions, by a process she does not at all comprehend, but evidently not by deductions of reasoning, but by a process originating in her mental and sensuous sympathy.

It is these characteristics which make female teachers more successful with young children.

But in the progress of development, when the Social and Animal Propensities of the children gain more force, especially in boys, male teachers are required. They ought, however, still to maintain the same sympathetic control as the female teachers.

The number of children in a family has an important relation to their training; because if the mother's attention is concentrated upon one or two only, the constant repetition of the exercise of the faculties which already predominate tends to special and unequal development, which results in angularity of character. This is why the *only* child is generally a *peculiar* child, and often a *spoiled* child. Where the family is large, but not too large for the care of the parents, the activities of the faculties of each child act and re-act upon those of the others, sometimes in sympathy, and sometimes in opposition. The mother supervises the children in these immediate relations with each other, and checks extreme manifestations, and allays the friction, but does not interfere except by necessity. In a large family, therefore, there is a greater scope for the free and equal development of the child, and greater breadth of capacity for future action is given, than if the child is isolated or nearly so, and under the more exclusive influence of the parent. These conditions the teacher should strive to preserve in the school, and should preside over the mutual relations of the children toward each other, with the same sympathetic supervision, and abstinence from unnecessary interference, which the mother maintains in the most favorable conditions of family education.

And as new tasks and burdens are to be imposed upon the child, when it leaves the mother to enter school, it is of the utmost importance that the teacher possess and conform to a knowledge of the laws which govern growth and development, especially of the physical system, in children.

Desiring in this letter to confine myself to the elucidation of a single subject, I shall refer only in a very limited and special manner to the laws of growth as they affect temperamental characteristics; but I will, in some future communication, trace their influence upon the phases of mental development comprised in the process of education.

You have already understood from me, in conversation, that the leading measure which I have thought might be advantageously adopted at this time, for the improvement of our Public Schools, consists in a classification of the pupils, based partially upon these characteristics and conditions. The present system of classification brings together twenty, fifty or even a hundred pupils of equal proficiency, into the same class room, and under the same exercises and treatment, but makes no allowance for differences of character, disposition, etc. Now, I am convinced that it is practicable to take notice of the distinctions of temperament at least, and in a way which will increase the efficiency and success of the teacher's efforts, and at the same time will not necessarily involve any change in the existing mode of classification. The principles I advocate may be applied to classes as they are now formed; though, if my views were fully and extensively carried out, and their application found as useful and successful, as I believe it would be, they would ultimately modify, somewhat, the selection of pupils for the various classes.

IMPORTANCE OF REGARDING TEMPERAMENTAL DIFFERENCES.

A few remarks upon the general principles on which the distinctions of Temperaments are based, will, perhaps, be not deemed irrelevant. Theoretic writers have suggested several systems, some of which, (for example, that of Powell, of which a good account is given in Appleton's "New American Cyclopedia"—article, *Temperaments*)—are carried into very minute subdivisions. It would be more correct to discard arbitrary divisions, and analyze the constitution of each individual as it is "tempered" by the peculiar admixture of the influence of the principal physiologic functions of the system; but for the purpose of a practical application to Education, the subject must be treated in a more generic method; and only what are commonly known as the four leading temperaments need be regarded. They are:

    The Nervous temperament,
    " Sanguine "
    " Lymphatic "
    " Bilious "

These are extensively understood, having been recognized by physiologists since the time of Hippocrates, and are well marked and easily distinguished, without requiring any special education on the part of the observer. The Principals of our Ward schools could easily qualify themselves to divide a class into four portions, according to the predominance of the four temperaments, respectively, and such a division would be thorough enough for practical purposes. In order to illustrate this, I will here enumerate the principal characteristics of each.

The peculiarities of the Nervous temperament spring from the fact that in such a physical organization, the brain and nervous system predominate, and their indications take precedence, in the make-up of the individual, in proportional size and in activity. The functions of mental life are stronger than others in the system. The Sanguine temperament, in like manner, indicates the predominance of the lungs and arterial system as compared with the other physiological functions. The Lymphatic temperament is accompanied by a similar predominance of the functions of the stomach and digestive apparatus, and of the glandular and lacteal system; and the Bilious, by similar predominance of the functions of the liver,—the great secreting organ of the body.

To appreciate how much the susceptibility of a child to mental impressions is modified by the temperamental character, it must be borne in mind, that the brain is the seat and centre of mental manifestations; the lungs, the seat of the warmth and forces which the atmosphere gives; the stomach and lymphatics, the organs of liquid supplies for the system; and the liver, the organ of secreting from those supplies the material for sustaining life.

Causes of Mental Vivacity, Persistency, Placidity, Sluggishness.

Three things are always to be taken into view, when estimating any of these bodily organs, its size, its quality, and its activity. Size and quality or degree of development, afford a measure of power; activity of influence. Each organ is to be estimated, not absolutely, by what it is in itself, but relatively, and with reference to the influence it may exert on, or receive from the other three functions, especially the brain, the organ of mental life. Thus, in a constitution in which the brain is the predominant organ, if the lungs are active, they give, by the higher arterialization of the blood, more warmth and more sensitiveness to the influence of the brain; and, where size in the lungs is super-added to activity, increased power and continuousness of influence are acquired by the brain. But when the lungs are of such activity and size that they form the predominant organ in the body, then their activity, by spasmodic warmth, tends to supersede that of the brain, and the latter in proportion loses the power of continuous effort. Again, in respect to the stomach, digestive tract, and lymphatics, size, especially when combined with activity, produces fulness in all the glands and tissues, and by maintaining a continuous supply of liquids and keeping the capillary vessels surcharged, restrains the action of the brain in respect to vigor and intenseness of mental effort, but gives ease and freeness. The mental action is no longer impulsive, but becomes placid. When, however, size and activity reach so extreme a point, that the stomach becomes the most influential organ of the body, sluggishness, indifference, and lassitude in the action of the brain and lungs are the result.

Again, when the liver possesses a due degree of size and activity, it gives tone to the brain, by maintaining a proper supply of the nutritive blood, ready to be aerated by the lungs, for the demands of the whole system. But if the liver is increased so as to become the preponderating organ, it overcharges the system, and the result is torpidity through want of vitalization. In proportion also, as the functions of the liver are active, the individual is disposed to withdraw from apparent activity, and to act in secluded relations. This disposition to retirement is connected with sleep and those quiescent conditions which are essential for the peculiar action of the liver, in secreting the necessary liquids ready for arterialization, and is strikingly analogous to what is known in animals as "hibernating." The functions of the liver have a relation to low temperature, as those of the lungs do to heat. The liver supplements the action of the lungs, by eliminating from the circulation elements which, if left to be consumed by the lungs, would produce excessive heat. By this function it preserves the conditions for equalized temperature.

In respect to all these, I speak only of the healthful state. Disease often modifies both size and activity.

It is the function of the Nervous system, of which the brain is the leading organ, to preside over the expenditure of the forces which these three lower temperaments afford. It can do no more, so far as physical

Peculiarities of Character, Mental and Physical, arising from Temperaments.

laws go, than to command and direct those forces, such as they are, in accordance with its organic mental nature; and the amount and qualitative character of the force depends upon the combination of these three temperaments, in their relation with the Nervous system.

The activities of the brain in children are either sensory and motor, or intellectual. These must be exercised in accordance with their Physiologic constitution. The whole system suffers if the brain and nervous system are over exerted in either of these parallel activities. And since, in children, the sensory motor force is, when under the unrestrained influence of their sensuous growing nature, too actively affected for receiving instruction, the special attention of the teachers should be devoted to an equalization of this excess by a discriminating expenditure in the intellectual powers, which depend upon the sensory and motor for their vital force.

Each of these four temperaments imparts its own peculiar character and expression to the face and head, and, in a less degree, to the whole system, not only in color, but in all the manifestations of life and activity. Thus, the liver, when active, imparts to the system tinges of black, and this expression is characteristic of the Bilious temperament. The lungs, when they predominate, give a brilliant red, imparting the fiery expression which characterizes the Sanguine temperament. When the stomach and lymphatics are the leading organs, a torpid, lead-like tinge of white is perceived, accompanied by an expression of lassitude in all the muscles and tissues. And the brain, when it leads in the organization, imparts vividness, quickness, and sharpness of movement in all the mental manifestations, and gives a clear whiteness and refinement to the whole expression, and a sharp outline to the features of the head.

In the case of children, growth being the leading necessity of life up to the age of puberty, the lymphatic conditions, as a general rule, predominate. Subject to this law, however, there are alternations of the other temperaments, from which arise very great and important differences or modifications of character.

Children of a Nervous temperament are quick in the action of the brain, and when the brain is well developed, are noticeable for intelligence and apprehension; they are, relatively speaking, eager to learn, and learn easily and fast, and are readily impressed through the mental faculties. But they are less able to retain what they learn, and are more easily diverted from the effort of learning, than those of the Bilious temperament; have less warmth of temper in all mental dispositions than the Sanguine; and are less susceptible to our ordinary methods of mental training than those of the Lymphatic temperament.

The Sanguine children are volatile, and more swayed by pleasures of the senses, and less by things which attract the mind, than the Nervous ones; are less persistent than the Bilious; require more tact and care in their education than the Lymphatic; but their superiority in warmth and active energy, arising from higher arterialization, renders all exercises and modes of education which involve the use of the physical organs, easy and attractive to them.

The children of a Lymphatic temperament are easily swayed and led by the will of the teacher, receiving *impressions*, as distinguished from ideas, easily. They will do as they are urged to do, willingly, but are slow of comprehension, as compared with the Nervous, and inert in respect to physical activity, as compared with the Sanguine, and changeable or variable in purpose and effort, and deficient in retaining impressions, as compared with the Bilious.

The Bilious temperament, on the other hand, gives permanence to all impressions, enabling a child to retain mental impressions when once acquired, though their original acquisition is generally more slow and difficult than in the case of the Nervous temperament. Such children, too, require to be dealt with in a more private way than others, the disposition to retirement being a striking trait of the temperament. This temperament relieves the child, in some measure, of the temptations which outdoor sports and amusements offer so powerfully to the Sanguine.

When we consider that children in a school are collected, not, as operatives in a factory, for what they can *do*, (for if that were the object, the proficiency of the individual might well be the sole ground of classification), but altogether for what can *be done* to them— what they can receive,— it is evident that differences of temperament, which involve such important variations in the proper mode of training, cannot be ignored in classification, without seriously affecting the results of education.

It may be objected to this, that the children of each temperament need the influence of other temperaments to modify or stimulate that which is excessive or deficient in themselves; and that the commingling of the children tends to equalize them, like mixing soils, quickening those that are slow, and steadying those that are too volatile. This is very true and this influence must be secured. I have previously pointed out the most substantial reasons why an aggregation of children affords more favorable conditions of development than isolation. But the error is in overlooking the distinction between the general arrangements upon which the intercourse of the children depends, and which is of importance to the development of the temperamental systems, and special arrangements for the mere purposes of imparting information and training particular faculties.

The careful observer of children will see that the intermingling which secures this equalizing influence is compatible with a classification by temperaments for the purposes of instruction. It is in the play ground and in the general relations of the school, that the influence of the Sanguine child will arouse and force into activity the Sanguine system of the Bilious, the Nervous, and the Lymphatic children, and that the mental superiority of the Nervous child will awaken the mental force and the ambition of the Sanguine, the Bilious, and the Lymphatic. But when the children come to the teacher to receive information, his labor is wastefully applied, if those who are quick and those who are slow,— those who remember upon the first statement, and those who must hear again and again, —are all commingled. The existence, in a class of slow minded pupils, of a section of quick minded ones, instead of having the effect to accelerate the acquisition of knowledge by the former, will be found generally to tend to confusion and superficiality by urging them forward, while it produces listlessness and inefficiency in the minds of those who are held back; and this necessarily deranges the whole, and causes great additional care and labor on the part of the teacher, and makes teaching more exhausting.

It is not, however, to be supposed that the mental disposition of the child resides in the temperament. This depends directly upon the organization of the brain; but the temperamental conditions exert a marked influence upon the activity of the brain, and, both directly by growth, and indirectly by the senses, modify the mental disposition.

An intelligent recognition of the laws of growth, and a sympathy with them, and an adaptation of the processes of Education to them, such as I desire to see exist, will give the teacher a conscious restraining and guiding influence, not only over the development of the mind, but also over that of the temperamental character.

To understand how to deal with either temperamental or mental peculiarities, the teacher must regard the vital sensibilities of the child, for it is upon and through these that we work.

These sensibilities are of two classes, 1, bodily or physical, and 2, mental.

The bodily sensibilities include those vegetative forces or processes which are of a nature common to the life of both animals and plants, giving support and supplying the waste of life, and which I have briefly described as the Sanguine, Lymphatic, and Bilious temperaments.

Through the Nervous system, the peculiar characteristic of animal life, the physical sensibilities of irritability and contraction are reported in the brain, causing thereby mental sensibility, either of pleasure or comfort, or of pain or discomfort.

Since childhood is a period of bodily growth, these physical sensibilities are very active and variable in children.

The first condition, therefore, for controlling the child's mind is attending to these bodily sensibilities, to avoid all excitements, either painful or pleasurable, which would tend to supersede mental sensibility during the period appropriated to instruction.

The time of instruction, the bodily positions, the exertions and the constraints of the scholars, the light, air, temperature, and all the conditions of physical want, should therefore be adapted with reference to their age, to secure bodily comfort.

Treatment of the Mental Sensibilities. Combined action of the Senses.

The mental sensibilities are either, 1, those of the Propensities, Social and Animal,—that is the Desires and restraints of desires, or 2, those of the Intellectual powers, or 3, those of the Moral and Spiritual feelings. For reasons I shall hereafter explain, the teacher deals less with the moral and spiritual sensibilities, than with the sensibilities of the body and organs of special sense, and those of the Propensities and the Intellect.

The sensibilities of the Propensities are awakened, primarily or immediately, by the bodily or physical sensibilities; but are also awakened and stimulated by the sensibilities of the Intellect, through the organs of special sense, which, in addition to these internal excitants, awaken perceptions of external objects to satiate the desires. The desires of the Propensities are thus made intelligent.

The sensibilities of the Intellect, in like manner, are awakened primarily by the organs of the special senses,—sight, hearing, touch, taste, smell,—and act in accordance with the demands of the Propensities, which constantly stimulate them. This reciprocal action makes up the continuousness of the natural exercise of the mind.

The organs of sense should have the teacher's especial attention. In no part of our organization is the Wisdom, Power, and Goodness of God so obvious to the minds, even of children, as in the design of these avenues of intelligence. By the eye, the mind receives only perceptions which light can convey, which are either of the intensity or the color, or both. By the ear, the mind receives only perceptions of sound, which are either of its intensity, its pitch or acuteness, and its quality of tone. By the skin, perceptions of temperature. By the muscular sense (which is often described as a part of the sense of touch), the perception of resistance. By the gustatory and olfactory nerves, respectively, perceptions of taste and smell.

These organs of special sense, may act upon the mind simultaneously, and much of the knowledge or mental conceptions which are commonly regarded as acquired through one sense, could not have been acquired except by the simultaneous operation of two or more.

For instance by the eye, in connection with the muscular sense, the mind gains perceptions of solid form, size, motion, distance. Perceptions of the same qualities may also be gained by the ear in connection with the muscular sense. To give full and accurate perceptions of these qualities, all these senses should be trained together. By the eye and ear together, aided by the muscular sense, the mind gains those acute perceptions of rhythm necessary for the Musical art. The knowledge of flowers is gained chiefly through the simultaneous action of the eye and the olfactory nerve. The knowledge of foods, through the simultaneous action of the eye and gustatory nerves, aided by the olfactory nerves.

Childhood is the period of activity in these functions; and the teacher should study them and their combinations, if he would successfully develop the sensibilities and instruct the mind.

HOW TO ACCOMPLISH CLASSIFICATION BY TEMPERAMENTS.

I have said that, for the present, the classification by temperaments may be entirely subordinate to the existing arrangements of the schools. The following are the steps by which the experiment should be tried:

I. The new classification should be introduced through the action of the principals and subordinate teachers employed in the present system. I have never contemplated that it should be dependent on the knowledge or skill possessed by a particular individual outside the organization of the schools, but that it should be administered within the schools, by the same persons who are intrusted with the usual management and instruction. At the outset, the experiment would necessarily involve and require special explanation and aid from some person who, like myself, has made the subject one of special study. Such explanation and aid could, however, be readily given to the principals and teachers, by printed descriptions, and the method should not be pressed faster nor further than they find that the result is satisfactory.

II. In initiating this method, my first effort would be to obtain the concurrence of the principal of some one of the primary schools, in the attempt. I should ascertain that she had a general knowledge of the leading temperaments, and the physical signs by which they may be known: there are, of course many, which, for want of space, are not alluded to in this letter. I should then ask her to arrange the pupils of each class into four divisions according to her own judgment of their temperaments, and that they should be so seated in the class-room, that the different temperaments would occupy separate places. This being done, I should visit the class from day to day, and confer with the principal and teachers upon the different modes by which these divisions might with advantage be managed. Upon this head, explanations cannot be given in detail upon paper, but, in the presence of any ordinary children, I should be able to make them at once. These Nervous children will understand their lessons quickly, but they will forget them; they must have reviews—the same thing over and over again. These Bilious ones will not understand easily; you must be patient and take plenty of time in explaining every thing fully, at first; but what they have once learned they will remember. These Sanguine, ruddy-faced boys by the window, are not the ones to sit where they can look out of doors; every thing they see in the street, while under instruction, will distract their attention. Put them there by the door, and let the full-faced, watery, Lymphatic boys, now sitting by the door, go over by the window.

III. When the experiment has been tried in a single class, the principal and teacher can determine for themselves whether to go further or not. If the results of the system prove it to be advantageous, the next step in the work will be to extend it to another class, and this being accomplished, to a third, and so on.

Suggestions for Classification by Temperaments, Continued.

IV. The arrangement of pupils according to temperaments, in the manner thus described, if carried out through the whole school, would render it possible to improve the organization still further, by assigning the different teachers to the classes in such a manner that the temperament of the teacher and the class would harmonize. The extent to which this would be practicable must depend upon the number of teachers and pupils in the particular school.

V. Whenever the proposed system shall have been found successful in any one of the schools, similar measures can be employed in introducing it into others, if desired.

This classification is as important in its influence on the teacher, as in itself directly, because it will facilitate his study of various dispositions, and his adaptation of himself to the character of his scholars.

In schools or classes too small to subdivide, and down to dealing with the individual scholar, the teacher who gives attention to the laws of growth, and studies the characteristic differences of disposition manifested in different phases of organization, will find the facts to which I have called attention, equally available and useful, by enabling him to understand his pupil and modulate his own method and bearing as the case may require.

In conclusion, I desire to express my thanks for the time and attention you and your associates have bestowed upon the consideration of my views; and

I remain,

Very Respectfully Yours,
JOHN HECKER.

[*From Mr. Hecker to Mr. Kiddle.*]

With the above letter the following note was sent to Mr. Kiddle.

NEW YORK, *June 3, 1865.*

HENRY KIDDLE, Esq.,

DEAR SIR:—Herewith, you will receive a communication from me, stating my views in regard to improving the classification of pupils in the Public Schools. If any points in it should require further explanation, I hope you will do me the favor of giving me an opportunity to afford it. A note from you, suggesting any further questions, would receive early attention from me, or it may be easier for you to note the questions occurring to your mind, in the blank margins, returning the same to me for answers to them. I suppose it is understood that, for the present, my letter will only be laid before your immediate associates and such other persons as you may think it important to consult with upon the views presented. Whether it be desirable to make the suggestions I have offered, public, will be a matter for future consideration.

Respectfully Yours,
JOHN HECKER.

Interrogatories, Physiological and Phrenological.

[*From Mr. Kiddle to Mr. Hecker.*]

JOHN HECKER, Esq.,

NEW YORK, July 27, 1865.

MY DEAR SIR:—I mislaid the brief note of points of inquiry, which I made while in conversation with you a short time ago. I have, however, in compliance with your request, made an effort to get my mind on the same track, and send the following interrogatories as the result. They are of course very crude from the cause which I mentioned in my last interview with you—the want of time, for an adequate consideration of the subject.

1. May not all the facts of phrenology and the distinctions founded thereon be considered physiological, seeing that they have their origin in peculiarities of physical organization?

2. If so, would not education if based upon it, take cognizance, as the foundation of its discriminations and adaptations, of exclusively physical peculiarities?

3. Would it not then, as a developing or training process, be based in its practical operations upon, 1. Peculiarities of temperament; 2. Peculiarities of cerebral structure?

Such being the case, the following questions would arise:

(A) AS TO TEMPERAMENTS:

1. Would the division of temperaments into the four primary classes be sufficiently minute as a basis, without taking into consideration the various combinations as they usually occur?

2. If combinations are to be considered, is the prevailing temperament in all cases, to be the guide?

3. How are these distinctions of temperament to be made available, 1. In discipline; 2. In instruction?

(a) *As to Discipline:*

1. What temperaments are best treated by coercive means?
2. What by persuasive?
3. What other considerations are applicable?

(b) *As to Instruction:*

1. What temperaments are most inclined to study?
2. What modifications in treatment should this lead to?
3. What temperaments need stimulating to study?
4. What considerations as to the *different kinds of study* have reference to the several temperaments?
5. What other considerations with regard to temperaments?

Phrenological Interrogatories, Continued.

(B) AS TO CEREBRAL STRUCTURE:

1. What general principles (if any), founded upon external manifestations of cerebral structure, may be adopted as a guide in training the faculties of the mind?
2. Where any organ, for instance, exists in excess, what would be the proper treatment?
3. What, in case of deficiency?
4. In what order should the faculties be trained?
5. What is the proper classification of the faculties with respect to education?
6. How may the perceptive faculties be trained?
7. What faculties, phrenologically speaking, may be regarded as conceptive?
8. How should they be addressed and trained?
9. What faculties are constructive?
10. What treatment is proper for them?
11. At what stage should the reasoning faculties be addressed and exercised?
12. What moral faculties claim an early attention?
13. How to be trained?
14. What other considerations have reference to this point, in such a general summary as the above?

You will perceive that these questions are very general, and perhaps you will consider some of them vague. If you will, however, reply to them, I may, perhaps, be able to ask others more minute and definite. Your request at this time has compelled me to present this synopsis somewhat prematurely, and, therefore, it is not as well considered as I designed to have it.

Very truly yours,
HENRY KIDDLE.

Importance of Foregoing Interrogatories.

[*From Mr. Hecker to Mr. Kiddle.*]

To the foregoing series of interrogatories, Mr. Hecker sent the following reply in part.

NEW YORK, *August*, 1865.

HENRY KIDDLE, Esq.,

DEAR SIR:—Your note of July 27th, in which you present a series of questions with respect to the applications of Phrenology and Physiology to Education, was duly received. My answer has been delayed by absence from the city. I commence, in this communication, to respond to the inquiries in your note.

Permit me, at the outset, to thank you for the time and attention bestowed upon my suggestions, and to express my sense of the completeness and breadth of view, by which the system of interrogations propounded by you is marked. Those interrogations cover a very broad and comprehensive field of educational science, and indicate a logical and lucid method of treating the subject. They form an exceedingly good outline for a somewhat extended development of true views upon education; and I shall be grateful to God if I am enabled to present the truth, in answer to your questions, in a manner as clear and lucid as the outline thus furnished me deserves.

At this point I will humbly ask to have some explanation and assurance from you as to the extent to which, through our past conversation and correspondence, my views upon education have appeared to you to be worthy of confidence, and calculated, if properly developed, to attain actual success. I feel that there is a need, in preparing my answers to your questions, that I should be guided by some knowledge of the impressions which my communications have thus far made upon your mind. The truths which I urge are, in many respects, difficult of conception; but of the greatest practical value, and the facts upon which they rest are easy of perception. These truths embrace a knowledge of the laws of our mental life, based upon observation, and a classification of the faculties into three groups, the Spiritual, the Intellectual, and the Social and Animal. This knowledge, which through many years of study, labor, and experience, and by the assistance of Almighty God, I have been enabled to acquire, I would gladly communicate to those whose mission it may be to give it a practical and useful application for the benefit of my fellow men.

Phrenologic Bust of Washington, and its Physiognomic Characteristics.

Throughout my discussion of the subjects touched upon by your questions, I shall have frequent occasion to refer to the new Phrenologic Bust, which I have had prepared for the purpose of exhibiting, more correctly than has hitherto been done, the significance of the general form of the head as an indication of mental disposition, and the locality and grouping of the convolutions corresponding to the faculties into which we analyze the activities of the mind. As a subject for this bust—desiring to have an actual, not an imaginary, head—I have taken the head of George Washington. The bust itself, which is of life-size, is modelled from the mask taken from the living head; and the contour of the upper and back part of the head is modelled according to the mental character which he manifested in actual life. The colossal bust in marble, which you have seen in my library, is made in that design. This bust represents George Washington at about the age of forty-five, the most active and nervous period of his life. It exhibits the marked predominance of the Lymphatic temperament, producing torpidity of the vascular system, by which George Washington was characterized, in the earlier and again in the later years of his life, and which, taken in connection with the large glandular features of the head and neck, imparted to his character its peculiar scope, while these appearances would have led the mere Physiologist and Physiognomist to pronounce the elements of a great character wanting. In it, also, are corrected the errors, or deficiencies, which are noticeable in every portrait or bust of the Father of his Country,—such as giving too youthful or too aged an expression to his countenance, and the failure to represent aright the top and back of the head, in which are embodied the most commanding elements of character. In this bust, a sufficiently youthful expression for the age represented is given in the front view, while the profile or side view exhibits the maturity and steadfastness of character by which George Washington was distinguished; and the preponderance of the Spiritual Faculties, which in him was quite extraordinary, is also exhibited.

It would seem to the Physiognomist, who deals merely with form and features of the body and face, that indications of tone, power, and strength were wanting in the *front* view. Hence Lavater, when he first saw the cast of Washington, which presented this facial aspect, declared it could not be a correct delineation, as it presented no evidences of the moral and intellectual characteristics of this great man. This accounts also for the indistinctness and similarity of expression which mark the portraits of different artists of the day. Both artists and Lavater were unacquainted with the effects produced by temperamental and surrounding influences, and deficient in the scientific knowledge of the mind, which requires a combination of the principles of Phrenology and Physiology. But the Phrenologist, directed by the present advanced state of science, who duly estimates the Phrenologic features of the head, and the temperamental Physiognomic character of the face, as presented in this front view,

Characteristics of this Bust of Washington as Compared with other Phrenologic Busts.

will appreciate the facts which constituted the elements of his greatness. Upon the plaster cast of this head are delineated, by raised letters, on one side of the head, the Phrenologic name and general location of each faculty individualized; and upon the other side, the three lobes of the hemisphere, forming the three groups in which these faculties are associated.

This bust differs from that usually sold by lecturers, in these respects:

First, It is a true representation of an actual head, exhibiting the phrenologic form, grouping, and development of a real character; while others are more representations of a generalized ideal of the individual faculties, not presenting their actual groups, by which, alone, the development can be properly analyzed and understood.

Second, The names employed to designate the faculties of the Spiritual group are made to correspond with the best view of their nature. The nomenclature heretofore generally in use was framed for the natural and moral qualities, and is not consistent with the order of the spiritual nature of man. Of the necessity of recognizing this, I shall have occasion to speak fully, when I come to describe them, in answer to your questions. I employ a nomenclature which expresses this recognition.

Thus instead of the names introduced by Dr. Spurzheim:—	I employ the following:—
Veneration.	Godliness.
Benevolence.	Brotherly-Kindness.
Firmness.	Steadfastness.
Conscientiousness.	Righteousness.
Hope.	Hopefulness.
Marvelousness.	Spiritual Insight.
Imitation.	Aptitude.

Third, Instead of numbering the organs continuously throughout, as in the ordinary mode, I have numbered the organs of each group, in separate series, designating the members of each series in the order in which they should stand when fully developed in the adult.

One bust of this sort is now completed in plaster; and I am to be furnished with copies as fast as they can be prepared. As soon as they are ready, I shall place one of them before you for examination.

In the diagrams, I present this bust in various points of view, to illustrate the method in which the head should be looked at, to form an intelligent estimate of its Phrenologic form.

The circular lines about the head are not the representation of a measuring apparatus, but merely a diagram of geometric form, to assist the eye in analysing the curves presented by the outline of the head, and to show at a glance whether the characteristic development is upward, forward, backward or downward, from the central base line passing through the opening of the ears.

ANSWERS TO THE QUESTIONS.

The first three questions put by you are connected in subject, and I shall now proceed to give answers to them:

I. "*May not all the Facts of Phrenology, and the Distinctions Founded thereon, be Considered Physiological, Seeing that they have their Origin in Peculiarities of Physical Organization?*"

I answer:—That, in addition to facts and distinctions which have their origin in peculiarities of physical organization, and which are, therefore, within the domain of Physiology, as now taught, Phrenology and Physiognomy, when rightly understood, recognize other facts and distinctions, which are beyond the domain of Physiology, but which must be taken into account, if we would understand the mental nature of man.

The principles of the Spiritual nature of man, originating in God the Father, and manifested by the Son, and imparted by the Holy Ghost to man, are not Physiological but Spiritual. When the Spirit of God is recognized by the soul, these Spiritual truths are spiritually discerned. "That which is born of the flesh is flesh; and that which is born of the Spirit is Spirit." "Marvel not," said our Lord Jesus Christ, "that I said unto thee, Ye must be born again."

These are Spiritual truths, and do not have their origin in peculiarities of physical organization, but are resident therein. Man is the temple of the Holy Ghost.

A recurrence to the established facts of Physiology and Anatomy, bearing on this subject, will indicate how far mental phenomena depend upon physical conditions, and will prepare the way for what I have to say of these Spiritual truths, upon which depend the right development and activity of the faculties, their proper classification, the influence of the temperaments, and the Phrenologic observation of character.

PHYSIOLOGICAL FACTS.

Dr. Dalton, (*Treatise on Human Physiology*), has with great clearness and precision observed, collected, and generalized the perceptive facts of the subject. He sets forth the facts which have been ascertained, by a Physiological method rigidly confined to physical phenomena, excluding that class of physical phenomena, which rest in external manifestations in unimpaired living subjects. It will be seen, when we come to describe these external manifestations, that they add much to our knowledge;

Dr. Dalton's Physiological Description of the Functions of the Nervous System.

and when we take into view the Spiritual truths of which I shall speak, it will be seen that many functions, which he describes as involuntary, have been and may be brought under the control of man, through the higher Spiritual consciousness, by the power which is given to him by the Spirit of God, the Holy Ghost.

The following passages from Dr. Dalton's work will indicate what is known, through that interior physiological method, of the relation of the Nervous system to the Mind.

"In entering upon the study of the Nervous system," he says, "we commence the examination of an entirely different order of phenomena from those which have thus far engaged our attention. Hitherto we have studied the physical and chemical actions taking place in the body, and constituting together the process of nutrition"; and * * * "we have found each organ and each tissue possessing certain properties and performing certain functions, of a physical or chemical nature, which belong exclusively to it, and are characteristic of its action.

"The functions of the Nervous system, however, are neither physical nor chemical in their nature. They do not correspond, in their mode of operation, with any known phenomena belonging to these two orders. The nervous system, on the contrary, acts only upon other organs, in some unexplained manner, so as to excite or modify the functions peculiar to them." (p. 365).

The special endowment by which a nerve acts and manifests its vitality is a peculiar one, inherent in the anatomical structure and constitution of the nervous tissue. It is manifested, in the foregoing experiments, by its effect upon the contractile muscles. But we shall hereafter see that this is, in reality, only one of its results, and that it shows itself, during life, by a variety of other influences. Thus it produces in one case, sensation; in another, muscular contraction; in another, increased or modified glandular activity; in another, alterations in the phenomena of the circulation. The force, however, which is exerted by a nerve in a state of activity, and which brings about these changes, is not directly appreciable in any way by the senses, and can be judged of only by its secondary effects. We understand enough of its mode of operation, to know that it is not identical with the forces of chemical affinity, of mechanical action, or of electricity.

"And yet, by acting upon the organs to which the nerves are distributed, it will finally produce phenomena of all these different kinds. By the intervention of the muscles, it results in mechanical action; and by its influence upon the glands and blood vessels, it causes chemical alterations in the animal fluids of the most important character."—(p. 395-6).

After describing the spinal cord, and its two functions of assisting the brain in the production of conscious sensation and voluntary motion, he thus describes the reflex action of the spinal cord, as it takes place in a healthy condition during life, and when there is no intervention of the will of the subject.

"This [reflex] action readily escapes notice, unless our attention be particularly directed to it, because the sensations which we are constantly receiving, and the many voluntary movements which are continually executed, serve naturally to mask those nervous phenomena which take place without our immediate knowledge, and over which we exert no voluntary control. Such phenomena, however, do constantly take place, and are of extreme physiological importance. If the surface of the skin, for example, be at any time unexpectedly brought in contact with a heated body, the injured part is often withdrawn by a rapid and convulsive movement, long before we feel the pain, or even fairly understand the cause of the involuntary act. If the body by any accident suddenly and unexpectedly loses its balance, the limbs are thrown into a position calculated to protect the exposed parts, and to break the fall, by a similar involuntary and instantaneous movement. The brain does not act in these cases, for there is no intentional character in the movement, nor even any complete consciousness of its object. Everything indicates that it is the immediate result of a simple reflex action of the spinal cord."—(p. 413-4).

Functions of Cerebral Ganglia,—Sensation, Consciousness, Volition.

If the power of the soul, when its full consciousness is awakened, were taken into account, which cannot well be done without examining external manifestations, it would be seen that the soul may and does have the power of control over these naturally automatic motions. That which is naturally involuntary may thus become subject to the voluntary. The power of the will in disease, the endurance of martyrs, some of the phenomena of trances and of what is called animal magnetism, are instances of this power.

The results stated by Dr. Dalton, in reference to the relative functions of the hemispheres and the ganglion of the annular protuberance, (still without reference to the modifications introduced by the awakening of the higher consciousness,) are as follows :

"The hemispheres, or the cerebral ganglia, constitute in the human subject about nine-tenths of the whole mass of the brain. Throughout their whole extent they are entirely destitute, as we have already mentioned, of both sensibility and excitability —(p. 419.) The hemispheres, furthermore, are not the seat of sensation or of volition, nor are they immediately essential to the continuance of life.—(p. 421.) The powers which have been lost, therefore, by destruction of the cerebral hemispheres, are altogether of a mental or intellectual character; that is, the power of comparing with each other different ideas, and of perceiving the proper relation between them."—(p. 422). "The collection of gray matter imbedded in the deeper portions of the tuber annulare occupies a situation near the central part of the brain, and lies directly in the course of the ascending fibres of the anterior and posterior columns of the cord. * * * The tuber annulare must be regarded as the ganglion by which impressions, conveyed inward through the nerves, are first converted into conscious sensations; and in which the voluntary impulses originate, which stimulate the muscles to contraction.

"We must carefully distinguish, however, in this respect, a simple sensation from the ideas to which it gives origin in the mind, and the mere act of volition from the train of thought which leads to it. Both these purely mental operations take place, as we have seen, in the cerebrum; for mere sensation and volition may exist independently of any intellectual [*i. e.* mental] action, as they may exist after the cerebrum has been destroyed. A sensation may be felt for example, without our having the power of thoroughly appreciating it, or of referring it to its proper source. This condition is often experienced in a state of deep sleep, when, the body being exposed to cold, or accidentally placed in a constrained position, we feel a sense of suffering without being able to understand its cause. We may even, under such circumstances, execute voluntary movements to escape the cause of annoyance; but these movements, not being directed by any active intelligence, fail of accomplishing their object. We therefore remain in a state of discomfort until, on awakening, the activity of the reason and judgment is restored, when the offending cause is at once removed.

"We distinguish, then, between the simple power of sensation, and the power of fully appreciating a sensitive impression and of drawing a conclusion from it. We distinguish also between the intellectual, [*i. e.* mental] process which leads us to decide upon a voluntary movement, and the act of volition itself. The former must precede, the latter must follow. The former takes place, so far as experiment can show, in the cerebral hemispheres; the latter in the ganglion of the tuber annulare."—(p. 438-9.)

The functions of the Nervous system, investigated in this purely physical aspect, without including the effects produced when the higher consciousness is evoked, are thus summed up :

"We have now, in studying the functions of various parts of the cerebro-spinal system, become familiar with three different kinds of reflex action.

"The first is that of the spinal cord. Here, there is no proper sensation, and no direct consciousness of the act which is performed. It is simply a nervous impression, coming from the integument, and transformed by the gray matter of the spinal cord into a motor impulse destined for the muscles. * * * Actions of this nature are termed, par excellence, *reflex* actions,

Dr. Draper's Description of Functions of Nerve Cells and Nerve Fibres.

"The second kind of reflex action takes place in the tuber annulare. Here the nervous impression, which is conveyed inward from the integument, instead of stopping at the spinal cord, passes onward to the tuber annulare, where it first gives rise to a conscious sensation; and this sensation is immediately followed by a voluntary act. Thus, if a crumb of bread fall into the larynx, the sensation produced by it excites the movement of coughing. The sensations of hunger and thirst excite a desire for food and drink. The sexual impulse acts in precisely the same manner; the perception of particular objects giving rise immediately to special desires of a sexual character.

"It will be observed, in these instances, that in the first place, the nervous sensation must be actually perceived, in order to produce its effect; and in the second place, that the action which follows, is wholly voluntary in character. But the most important peculiarity, in this respect, is that the voluntary impulse follows *directly* upon the receipt of the sensation. There is no intermediate reasoning or intellectual process."—(p. 443.)

"All actions of this nature are termed *instinctive*. They are voluntary in character, but are performed blindly; that is, without any idea of the ultimate object to be accomplished by them, and simply in consequence of the receipt of a particular sensation. Accordingly experience, judgment, and adaptation have nothing to do with these actions." * * * *

"The third kind of reflex action requires the co-operation of the hemispheres. Here, the nervous impression is not only conveyed to the tuber annulare and converted into a sensation, but, still following upward the course of the fibres to the cerebrum, it there gives rise to a special train of ideas. We understand then the external source of the sensation, the manner in which it is calculated to affect us, and how by our actions we may turn it to our advantage or otherwise. The action which follows, therefore, in these cases, is not simply voluntary, but *reasonable*. It does not depend directly upon the external sensation, but upon an intellectual [*i. e.* mental] process which intervenes between the sensation and the volition. These actions are distinguished, accordingly, by a character of definite contrivance, and a conscious adaptation of means to ends; characteristics which do not belong to any other operations of the nervous system.—(p. 444.)

Dr. Draper, *(Human Physiology, Statical and Dynamical)*, describes in greater detail the structure and relations of the nerve fibres, or white matter, and of the vesicles, or gray matter; and infers that the functions of the nerve cells or vesicles are:—

"1. To permit the escape of an entering [influence out of the solitary channel in which it has been isolated into any number of diverging tracts; 2d. To combine influences which are entering it from various directions into a common or new result; 3d. By permitting of lateral diffusion to take off and keep in store for a certain duration a part of the passing influence."

"The registering ganglia thus introduce the element of time into the action of the nervous mechanism. The impression which without them would have forthwith ended in action, is delayed for a season, nay, perhaps even as long, though it may be in a declining way, as the structure itself endures; and with the introduction of this condition of duration come all those important effects which ensue from the various action of many received impressions, old and new, upon one another."—(p. 269.)

You are perhaps familiar with his profound and beautiful argument respecting the existence of the soul, as the agent necessary to act upon the "influential" apparatus which the hemispheres constitute, in order to produce the phenomena of human life, which the automatic mechanism of the simple registering ganglia could not produce.

"The introduction of a registering ganglion necessarily gives rise to an extension of the physical relations of an animal by connecting its present existence with antecedent facts, for the ganglion at any moment contains the relics of all the impressions that have been made on it up to that time, and these exert their influence on any action it is about to set up. In virtue of them, the nervous mechanism has now the power of modifying whatever impressions may be made on its centripetal nerves, and, within certain limits, of converting them into different results. Yet still the automatic condition is none the less distinct, and still the immediate source of every action is to be found in external impressions.

"An increasing complexity of nervous structure is next evidenced by a division of the registering ganglion into two portions, which, with some incorrectness, may be designated sensory and motor lobes, a division which is preparatory to, and, indeed, obviously connected with the introduction of a totally new method of action and source of power."—(p. 282.)

"The simple cellated nervous arc consists essentially of these portions, a centripetal fibre, a vesicle, and a centrifugal fibre; the centripetal fibre may have at its outward or receiving extremity vesicular or cellular material. Thus constituted, this mechanism is ready to receive external impressions, which, if such language may be appropriately used, are converted or reflected in part by the ganglion into motions, and the residue retained. But the arc, viewed by itself, is a mere instrument, ready, it is true, for action, but possessing no interior power of its own. It is as automatic as any mechanical contrivance in which, before a given motion can be made, a certain spring must be touched.

"The essential condition of the activity of such a nervous arc is therefore the presence and influence of an external agent—a something which can commence the primitive impression, for without it the mechanism can display no kind of result. Moreover, there must be an adaptation between the nature of that agent and the structure thus brought in relation with it, as is strikingly illustrated by each of the organs of sense."—(p. 283-4.)

"The problem we are dealing with * * * * may be stated, Given the structure of the cerebrum, to determine the nature of the agent that sets it in action. And herein the fact which chiefly guides us is the absolute analogy in construction between the elementary arrangement of the cerebrum and any other nervous arc. In it we plainly recognize the centripetal and centrifugal fibres, and their convergence to the sensory ganglia, the corpus striatum and optic thalamus; we notice the vesicular material of their external periphery as presented in the convolutions of the human brain; and if in other nervous arcs the structure is merely automatic, and can display no phenomena of itself, but requires the influence of an external agent—if the optical apparatus be inert and without value save under the influences of light—if the auditory apparatus yields no result save under the impressions of sound—since there is between these structures and the elementary structure of the cerebrum a perfect analogy, we are entitled to come to the same conclusion in this instance as in those, and, asserting the absolute inertness of the cerebral structure in itself, to impute the phenomena it displays to an agent as perfectly external to the body and as independent of it as are light and sound, and that agent is the soul."—(p. 285.)

In defining this nervous arc, as has been so clearly done in accordance with Physiology and Anatomy, at this point it would be profitable to observe how the mind in its three-fold nature acts upon it.

When the will-power is in the Propensities, and has command over the nervous arc, by its centripetal or sensory, and divergent or motor tracts, the will, which is put into execution through these tracts, acts in accordance with its own conclusions; whilst the Intellectual consciousness knows analytically the sensuous state of its own being, through the nerves of sense.

The Spiritual illuminating will of God, working not mediately but directly, supersedes the individual will, under intelligent analytic conditions, and by the special and additional higher consciousness thus given to the mind, subordinates the will of the individual. The organic manifestation of this is the Spiritual group of faculties marked on the bust. The Divine Will commands rather than disturbs, by its meek, humble, quiescent, and equalizing influences, when the subject is under the direct influence of the Godly Spirit, with the seven co-ordinated faculties exercised thereby, viz:—Godliness, Brotherly-Kindness, Steadfastness, Righteousness, Hopefulness, Spiritual Insight, and Aptitude.

Dr. Draper's Description of Ganglia and their relation to the Hemispheres.

Phrenology carries the investigation of the problem, which Dr. Draper has stated, forward towards its consummation, by comparing the varying form and appearance of the cerebral instrument with the functions performed by it in mental life, thus enabling us to analyze objectively the various activities of the soul. But our understanding of the facts thus gathered cannot be complete, without recognizing the influence of the Holy Spirit upon the soul, by which it is illuminated and the higher faculties of the nature are quickened.

I will merely quote in conclusion of these physiological statements, what Dr. Draper says, by way of description of the ganglia at the base of the brain, and their relation to the hemispheres.

"The ganglia at the base of the brain are regarded by Dr. Carpenter as constituting the true sensorium, a doctrine which he has established by many weighty arguments, and which is doubtless one of the most important thus far introduced by any physiologist.

"The idea here intended to be conveyed is, that the thalami, striata, sensory ganglia, and nervous arrangements below, constitute an isolated apparatus; distinct from which, and superadded, are the cerebral hemispheres.

"From observations on the animal series, the conclusion seems to be unavoidable that the chain of ganglia now under consideration must constitute a sensorium, the centripetal fibres communicating their impression and motion ensuing, the impression being attended with consciousness." * * * *

"But after the cerebral hemispheres are added, an impression received upon the thalamus, whether it has come in through the sensory ganglia, or any other sensory part of the craniospinal axis, is transmitted to the convolutions along the radiating fibres. From the convolutions, the influence which is to produce motion descends along the converging fibres to the striatum, thence along the inferior layers of the crus, through the mesocephalon to the anterior pyramids, and by their decussation to the opposite side of the cord.

"Such is the view which Dr. Carpenter presents of the functions of the sensory ganglia and spinal axis; or, employing the terms we have previously defined, the cord alone is a longitudinal series of automatic arcs; on the addition of the thalamus and striatum, it becomes a compound registering arc, the cerebral hemispheres finally annexed to it constituting an influential arc.

"In a simple arc, an impression is at once converted into motion, and leaves behind it no traces; its expenditure is instantaneous and complete. In a registering arc, a part of the impression is stored up or remains—nay, even the whole of it may be so received and retained. It is not to be overlooked that, as soon as the effect occurs, the evidences of sensation arise; and since sensation necessarily implies the existence of ideas, ideas themselves are doubtless dependent on this partial retention or registry of impressions."—(p. 319-21.)

From what is said above, we infer that Dr. Draper is not to be understood, here or in what follows, as asserting that consciousness or ideas have their seat in the thalamus; but that the purely automatic impressions or sensations of the sensory tract have an immediate relation to consciousness and ideas, which are manifested by the cerebrum, when, to use his figure, the soul touches the spring of this mechanism.

"There can be no doubt that the cerebral hemispheres constitute the instrument through which the mind exerts its influences on the body."

"From this point of view we may therefore consider the intellectual [*i. e.* mental] principle as possessing powers, properties, and faculties of its own; as being acted on by impressions existing in the thalamus, and delivered through the intervening fibrous structures to the vesicular material of the convolutions of the cerebral hemisphere. In this region they act upon the intellectual [mental] principle and are acted upon by it, the returning influence, if any, coming down through the converging tubular structures to the corpus striatum, and by its commissural connections sent off to particular ganglia." * * * (p. 321.)

In Mr. George Combe's Introduction to his translation of the views of Dr. Gall and others upon the *Functions of the Cerebellum*, (Edin. 1838) you will find a very clear summary of these facts, as foreshadowed by the conceptions and investigations of Gall and Spurzheim; and he points out the relation between the sensory tract and the posterior and middle lobes of the cerebrum, and that between the anterior lobes and the motor tract. I have preferred to cite our own later Physiologists, not because they were the first to discern these conditions, nor because they have carried the investigation in this aspect much beyond the general position marked out by Dr. Gall: but because they are familiar and accepted authorities upon the subject; and *as far as they go*, both with the delineation of perceptive facts of anatomy, the generalizations of physiology, and the resulting deductions of argument, they accord remarkably with the general principles upon which Dr. Gall's system was based.

Dr. Draper and Dr. Dalton both concur in the general opinion that we have no reason for denying that different parts of the brain may be occupied by different mental faculties. But the indisposition of Dr. Dalton to advance theories or principles, except as they are clearly the inevitable deduction from the facts he has examined, has led him to express the opinion that it may not be practicable for an observer to gather the facts requisite to establish Phrenology.

The other two objections stated by him,—*viz.*, that the layer of gray matter is continuous, without anatomical division of its interior structure, and that only a small portion of this surface can be examined by external manipulation, indicate that he was passing judgment rather upon the popular phrenologists of the present day, than upon the scientific principles advanced by Dr. Gall. Phrenology, properly understood, gives due significance to these facts.

It appears thus, that Anatomy and Physiology, as ordinarily understood, take cognizance of exclusively physical peculiarities, and do not detect the principle of life further than to describe the outgrowth of it in its orderly physical form; yet they do show us that there is a presiding force different from and superior to the chemical and mechanical forces known to physical science. One branch of the investigations of Phrenologists is a contribution to our knowledge of the mind, drawn from those external manifestations which Physiology, as ordinarily understood, ignores.

SPIRITUAL TRUTHS.

When all the physical facts thus gathered are examined in the light of the history and conduct of man, and the phenomena of consciousness, we see that they all point to the great paramount fact that the human soul in its oneness presides over this complex organization, for which it is responsible. The soul, although its manifestations reside in physical con-

-ditions, exhibits phenomena that physical laws and physical forces cannot explain. A study of the temperament and cerebral organization, is only an investigation of the external manifestations of this Spiritual existence.

The knowledge of this realm of Spiritual truth is not, like physical truths, acquired by Perceptive, Analytic or Synthetic powers, it is only possessed by *receiving* it. Nor is it to be communicated by language, as sensuously and ordinarily understood. Language can define what it is *not*, and can describe the phenomenal manifestations; but that which is Spiritual is only to be spiritually received, and spiritually discerned. The Spirit clothes His behests in a language written in the soul alone, and His power thus given, manifested through the Spiritual faculties, in varying phases of gifts according to the organization, communicates with, illumines, and commands the faculties of the intellect, the senses, the natural desires, and the temperamental conditions of all those who are the subjects of His influence.

In the experience of the individual, when these Spiritual laws are merely intellectually conceived and discriminated, or when their operation is made servient to the social and selfish life, the mind is centralized in the physical and sensuous nature, having no higher consciousness than a passional or intellectual force, modified by the suave influence of the sentiments: but when the mind submits to the Divine Power, and follows His direction in accordance with the principles of Christianity, man becomes a living soul, the instrument of God, who then " worketh in him to will and to do of His good pleasure."

Exhaustion of pride and selfishness, self-negation, both of the Propensities and the Intellect, and self-examination, not by the Intellect, but by the Spirit of God searching the heart, are the conditions of Spiritual wisdom. The intellectual man must cease to limit himself to intellection or logical proof, and the man of social or selfish passions, cease to engross himself in his dependence on the objects of his desire, and the soul must use those powers as the instruments of the Holy Spirit. Such is the scientific proof of the spirit of the Christian life.

The Spirit, when thus possessed, manifests Himself phenomenally through the physical organization, and His presence depends so far on cerebral conditions as to require that meek and humble state of preparation which arises from self-abasement and contrition. This meekness is one of the general conditions of the Spiritual group of faculties, a cerebral organization which all men who are morally accountable possess. Meekness arises either where there is predominant Spiritual organization, or from cultivation of these faculties, or may be induced by the instrumentality of external circumstances negativing the will of the Propensities and Intellect. The Spirit makes use of the mind when He is sought in this state of meekness.

The diversity of special gifts is in accordance with organization. The action of physical life being sensuous in its nature, is dependent upon temperamental and physical conditions for activity and qualitative character, but the phenomenal manifestations of Spiritual life, being Spiritual in their nature and essence, originate and receive their strength directly from God.

It does not follow, however, from the fact that the special gifts depend in part upon organic conditions, that large spiritual faculties in the natural mind are favorable to religious impressibility. For, while unawakened, they may and commonly do tend to spiritual pride, which is among educated men a greater obstacle to conversion than sensuous influences.

It is not strange that these truths relating to the mind have not been recognized by those who know no other method of investigation than observation of perceptive facts and the resulting processes of inductive reasoning. Perceptive facts verify these truths, when once placed, in a proper order of anatomical truth, before the mind.

It has been the misfortune of many persons who have espoused Phrenology, and have ardently advocated its claims as a science, that they have not been guided to recognize and develop the truth of the spiritual life of man. They have seen in him only the animal and intellectual faculties, and have supposed that all the phenomena of his nature and modes of action could be explained as physical phenomena. They have not admitted that there is such a spiritual condition as the Light of life; because they have not looked at man in his historical life, in which we cannot account for civilization and progress, except upon other conditions than those exhibited by the heathen world. Hence Phrenologists have ignored the spiritual nature of what they have called his moral faculties. They have recognized only the vitalized brain, and its manifestations when led by the Propensities or directed by the Intellect, and, in consequence, have overlooked the spiritual influences which operate upon man's nature and actions, when he is properly awakened to them, and have only ascertained the physiological laws of his life.

The early discoverers of the fact that the brain is the organ of mental manifestations, and that the convolutions of the hemispheres correspond to certain elements of mental power, in their desire to develop these most important truths, overlooked or neglected these spiritual laws, which are no less important; and their followers have, to a great extent, limited their labors to the task of testing, expounding, and advancing the physical part of our knowledge alone, instead of combining with it an exposition of the spiritual life. This combination must be made; a recognition of the distinct spiritual nature of man, and of his relations with his Maker, Almighty God, with the Grace of the Holy Spirit, and the example of our Lord Jesus Christ, must be associated with a true view of his physical organization, before Phrenology can take its proper place as a science, or perform its intended work as a practical and useful art. Both Gall and Spurzheim failed to accomplish this combination.

Dr. Gall's views are correct, to the extent to which he proceeded, and he foreshadowed, in some degree, the deductions which I have traced relating to the Spiritual nature; but his progress in the development of the science was limited by his moderate perceptive powers. For his character has been mis-conceived; since, while it has been stated that he was remarkable for power of the Perceptive faculties, he was not distinguished in that respect. When he became convinced that the memory of words resides in the brain, he conceived that this must have an organic manifestation consequent upon localized function, and accordingly explored mental manifestations seeking for the organic conditions of all peculiar mental phenomena.

This course was proper, so far as the Intellectual faculties and their manifestations are concerned. But the knowledge of God is not thus attainable. All attempts to establish a knowledge of the Deity by Intellectual processes, that is to say, by the faculties of the Perceptive, or Conceptive, or Combinative cluster, have failed to give a clear or intelligent faith, from the very conditions of our organization. He is only to be known by the faculties of the Spiritual group, that is to say, the Intuitive and Meditative clusters, of which Godliness or Veneration is the centre. These faculties are not to be characterized as logical or demonstrative, but as reflective or receptive, and in this is seen their special adaptation for giving us the knowledge of God by the presence and influence of the Holy Ghost. These faculties, too, led by meekness and Godliness, enable the mind to introspect its own Consciousness, and with the aid of the Intellect, analyze subjective phenomena.

The strength of Dr. Gall's mind, and the characteristics which enabled him to conceive and establish the elementary principles of Phrenology, lay in the philosophic Conceptive Faculties; which, with his large restraining faculties, gave him an intellectual forecast or forethought, sometimes characterized as deductive reasoning. If he had sought the truth through the Spiritual faculties, as well as by the Conceptive power of the Intellect, he would have included the necessary recognition of the Spiritual nature of man; while on the other hand, larger Perceptive powers than he possessed, in combination with his philosophic powers, would have led him to include among the facts which he sought for in support of his conception, the phenomena of the history of man, and the paramount influence which Religions, whether true or false, have always exerted in the events of that history. As it was, however, in the controversy in which he immediately became involved, he confined himself to the facts of physical organization, and did not enter into the higher applications of these principles to Spiritual life.

Dr. Spurzheim designed to give Phrenology the place of a science, and he gathered and presented the facts in a way that greatly advanced the investigation; but his classification is not in accordance with the attributes of man as a Spiritual being, for it is based upon the natural and sensuous phases of character common to man and other animals, and upon an arbitrary numerical order of the faculties; and therefore the subject had to encounter all the prejudices of established religious and scientific theories. He made, however, much progress in the anatomical investigations necessary to complete our knowledge of the subject; and had he lived to continue these inquiries, we cannot doubt that he would have been led to a new classification of the faculties. As it was, he arranged them according to his own cast of mind, or mental predominance, by intellectual discrimination, made entirely by the Perceptive, Conceptive, and Combinative faculties; and the basis of his classification is in the ideas he formed of the qualitative character of the respective faculties, as he had individualized them, and as he saw them manifested in men about him in worldly life, overlooking the modifications which Religion, in the true order of the Spiritual faculties, presents. Since, also, the mind, in the fallen state of man, tends to unequal and angular development, and to fragmentary conclusions, it resulted from the same characteristics of his method of observation, that he individualized the faculties more than is warranted by their actual grouping and continued action, and the general and outward historical facts, and did not sufficiently regard their activity as necessarily more or less composite, or associated, in the fallen state of man, and as unitedly and harmoniously associated, when the Holy Spirit guides and rules the whole soul, quickening the Spiritual faculties. Hence, he was led to give prominence in his system, to what is termed organology; an analysis, which, however much it may have promoted our knowledge of mental processes, has induced many scientific and logical minds to reject Phrenology; because the parts of the brain pointed out as distinct organs, are parts of a whole, and marked only by an orderly and limited difference of conformation.

From the observations which were the basis of Dr. Spurzheim's classification, he was led to deny that man can know God, and to characterize the Spiritual faculties as blind sentiments, and even to include in his enumeration of those sentiments several which are in fact a part of the Intellectual group.

When we unfold one of the hemispheres, and examine the organic position of the different convolutions, the functions of which external observation has shown, we find that they are arranged in a fixed order of internal contiguity, in a single series, the largest convolutions, being seven in number, forming the middle portion of the unfolded band, the medium sized convolutions forming one end of the band, and the smallest convolutions forming the other end. The central and largest convolutions are those which external observation of development has ascertained to be the instruments of spiritual or moral qualities, and, in their full exercise, they are found to correspond with the spiritual gifts described in the Holy Scriptures. This lobe of the brain is found to have but little sensuous communication, except through the adjoining lobes or ends, which form the other two groups we have described. The medium-sized convolutions upon one end, (forming the seat of the Propensities, both animal and intellectual), which, when enfolded in the encephalon, occupy the back part and lower sides of the head, and the smallest sized convolutions upon the other end, (forming the seat of the Intellect), which, when so enfolded, occupy the front part of the head, each have sensuous connections through and by the nervous system.

These facts indicate the existing classification of faculties, as ordained by the Creator in the organization of the brain. The manifestation and influence of these three groups of faculties, the Social and Animal, the Intellectual, and the Spiritual, seen in the history of Man, under Barbarism, Civilization, and Christianity—the characteristic power and function of the Spiritual group as delineated in the Scriptures, and realized in the consciousness of the Christian—and the contrast between men in worldly society, and the requirements of Scripture and example of Christ,—confirm my classification, and give profound significance to it. When we inquire what, in mental analysis, is depravity, and what are the new birth and the influence of the Holy Spirit, the central facts of personal religion, we see that the true life of man requires the predominant activity of these central faculties made sentient by the power of God, the Holy Ghost, upon the soul, and that His will, thus manifested in them, should subordinate the Intellect and the Propensities to His Grace.

But as I have said, the labors of both Gall and Spurzheim were limited to the task of developing Phrenology as a physical science. The fact of the correspondence of the organs of the brain with the faculties of the mind engrossed their attention; and the development, in detail, of the important laws of spiritual life, which should always be considered in connection with the physical facts, (many of which were first suggested by Dr. Gall,) was left for future writers.

I say, then, that the facts of which Phrenology should take note are partly physiological, and partly spiritual; that is, some have their origin in peculiarities of physical organization; and others in certain conditions of spiritual life already referred to; and that both these realms of truth, the physical and the spiritual, must be fully explored by all who would wisely and successfully apply Phrenology to the art of Education. In our presentation of the subject in its spiritual as well as physical aspect, it claims its true place as a science.

II. "*If so, would not Education if Based upon it, Take Cognizance, as the Foundation of its Discriminations and Adaptations, of Exclusively Physical Peculiarities?*"

This question has been already answered by the preceding remarks. A true system of education does not confine itself to the consideration of physical peculiarities. It studies these with care, and is guided by them in the larger number of its processes and modes of operation. But it recognizes the greater truth that man is not merely a living body, but that he is a spirit acting through an organized body; and it seeks to inform itself of his spiritual, as well as physical nature, and to deal with both natures by processes appropriate to each.

There is, it is true, a large field of labor in education, in which physical laws are the principal guides. But there is also a field of labor—also processes—in which spiritual truths *must* be recognized and followed, as the most important.

CHARACTERISTIC DISTINCTIONS AMONG FACULTIES.

Among the general principles relative to the qualitative character of different parts of the hemispheres of the brain, there are several which should be stated before I proceed to speak of the distinctions in the faculties with reference to physical and spiritual laws, and of Temperamental differences.

These I shall present more fully in answer to the inquiries in the second part of your letter. I only state them briefly here.

1. Each hemisphere is composed of three lobes or groups of convolutions, distinguished from each other, both by anatomical evidence of their sensuous connections, and by their contrasted functions. They are the posterior, the anterior, and the upper group, manifesting, respectively, the passional faculties of the mind, which are termed Propensities; the intelligent faculties, which are termed the Intellect, and the sentient or moral faculties, which I term the Spiritual group.

2. The nerves of sensation from the posterior column of the spinal cord and the thalamus more immediately communicate with the posterior group or Propensities; and here the natural and sensuous forces of the mind have their seat. The Intellect or anterior group is called into action both by the Propensities and by the nerves of special sense; and the passional desires of the mind, both selfish and social, thus become intelligent. The Intellect, by presenting and individualizing external objects of desire or necessity, re-acts upon and stimulates the Propensities.

3. The upper lobes, lying together along the median line in the crown of the head, have anatomically less immediate and full communication with the sensory and motor tracts than either of the other lobes; and these are the seat of the higher functions of religious impressibility and susceptibility.

4. The part of the hemispheres having more immediate relation to nutrition and voluntary motion are the convolutions contained in a central transverse core from ear to ear. This part comprises three exterior convolutions upon each side,—*viz*:—Alimentiveness in front of the ears, Destructiveness between or above the ears, and Combativeness behind the ears,—and also the interior convolutions of the Desire to Live. This

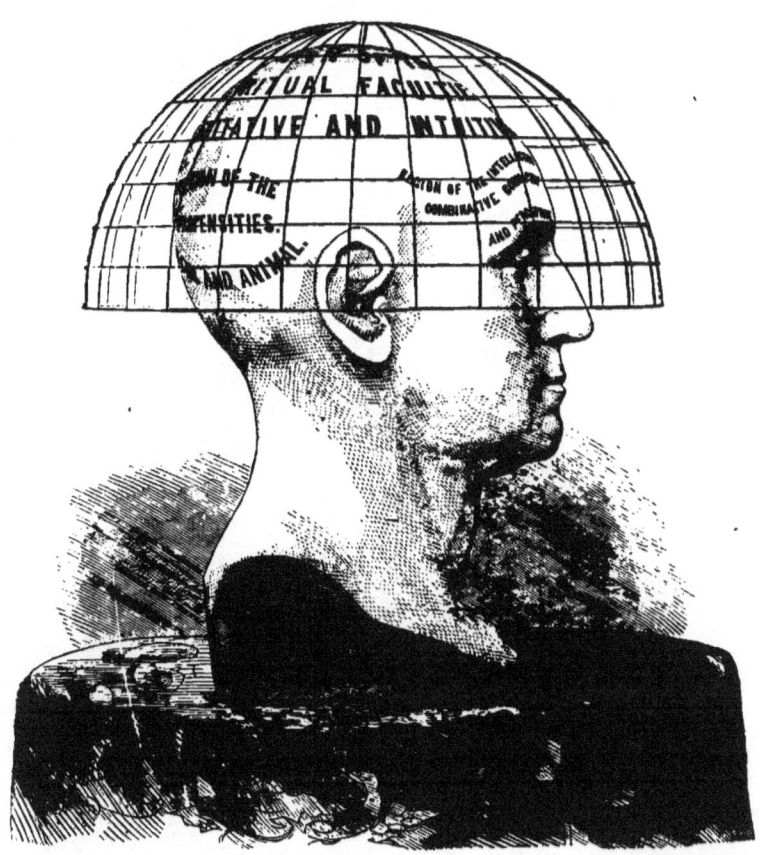

Photographic View (cut in wood) of the Phrenologic Bust of Washington.—*No. 1.—Right side.*—General localization of the three Groups. The largest letters representing the Spiritual Faculties, the next in size the Animal, and the smallest the Intellectual Faculties. The organs of the Propensities are in the back and side of the head, extending forward to and in front of the ear. The Intellectual faculties are in the region of the forehead and temples. The Spiritual faculties, in the coronal region, above both the other groups.

The curved lines around the head in this and the following views, are merely a diagram to aid the eye in recognizing the characteristic upward and backward direction of Washington's cerebral development.—See pp. 16, 117, 118. [To face p. 30.]

The Faculties of Restraint. Intercommunications of the Faculties.

part of the brain, especially the convolutions of Destructiveness and the Desire to Live, are the convolutions most closely connected with the medulla oblongata, and are the executive faculties, through which physical force of outward manifestation is given to the qualities of other parts.

5. The part of the hemispheres having more immediate relation to passional mental restraint, or the control of action and the retention of power, voluntary or involuntary, are the convolutions occupying a region above and behind the ears. This range comprises the faculties of Cautiousness and Secretiveness; there are two other restraining faculties, which being of a Spiritual nature are properly influenced only through Godliness. The two first named are among the Propensities, and the others, lying vertically above them, in the top and back of the head, are the posterior convolutions of the Spiritual group. These four pairs of faculties, I term the Restraining Faculties.

For the purposes of the present inquiries, it is not perhaps necessary that I should develop the anatomical description of the numerous, distinct convolutions which Drs. Gall and Spurzheim designated as the instruments or organs of some of the mental functions I am about to describe. But I may say, that by careful anatomical dissections, conducted under my own eye, I have become convinced of the existence of the individualized organs, as such, in the composite or unified form which constitutes the brain.

I cannot better indicate to the reader the general manner in which these organs of the brain are connected with each other, than in the words of Dr. Spurzheim (Anat. of Brain, p. 188.) "It is extremely interesting to trace the connection of the different cerebral masses composing special instruments. These connections explain the mutual influence of the faculties. The organs of analogous powers are regularly in each others vicinity; the convolutions that compose them even run into each other. The organ of Philoprogenitiveness communicates with that of Inhabitiveness, and with that of courage (Combativeness); the organ of courage communicates with that of attachment (Adhesiveness); and with that of Destructiveness. The organ of Secretiveness communicates immediately with that of Destructiveness, and with that of circumspection (Caution); the organ of Benevolence communicates with the organ of Veneration (Reverence); the organ of Firmness is in communication with those of all the faculties around it—Veneration, Justice, and Self-Esteem; the organ of justice runs into that of the love of acquiring (Acquisition); this is connected with that of construction (Constructiveness); the organs of the Perceptive Faculties are all linked together as are those of the reflective powers in like manner; the organ of artificial language is placed across the organs of the Intellectual Faculties generally. Thus the especial Design which God has taken to establish communications between the cerebral parts cannot be overlooked; and as I have already said, it is this arrangement that enables us to understand the mutual influence of their respective functions."

Fundamental Distinction between the three Groups of Faculties.

Let me now draw the line of distinction between the different modes in which the human faculties act, and indicate the distinctions of treatment which these modes require.

The law by which the faculties act in groups is of fundamental importance to intellectual education; and a full and constant attention to it enables us to correct tendency toward error, observable in the teachings, not only of many phrenologists, but of other intellectual philosophers.

The first step toward understanding the activities of the faculties is to know that by nature in the fallen State of man, their force resides in the Social and Animal Propensities, which are situated in the base of the back part of the head, within and above the occipital bone and the lower part of the parietal bones; while the Intelligence resides in the Intellectual Faculties, beneath the frontal bone, in the front part of the head; and the moral qualities and Spiritual disposition, which are the higher nature, reside in the upper part of the head, within the parietal bones.

The Propensities have an predetermined passional activity of desire, stimulated directly by the general organic sensibility, and indirectly through the organs of special sense, which communicate with the Intellectual faculties. This phase of mental activity is the universal law in children; and therefore the desires of the Propensities must be regulated and directed by education. The Intellect depends for its incipient action and stimulus on the outward influence of the senses, which act through the Perceptive faculties, and on the inward, conscious desires of the Propensities; and, therefore must be cultivated and instructed. The Spiritual Faculties being naturally dormant, having inherited the sleep from Adam, must be awakened; for as in Adam all men died, in Christ they all shall be made alive. This constitutes the phenomenal aspect of the individuality of the soul, by changing the predominant action from the Propensities to the Spiritual faculties which constitute the being man, a living soul.

The faculties have been individualized too much; the mind has been delineated as if it were a subject on the dissecting table, to be separated into parts, and studied only in its distinct functions; whereas, the faculties ought to be viewed, chiefly, as constituting a living whole, while the phenomena of their activity should be studied as composite organic acts. Herein, the methods of Dr. Gall were superior to those of Dr. Spurzheim. In all my views upon this subject, I am guided by the predominant fact of the association of the organs into groups; and I would caution all against the error of studying individual activity, without attention to the individuality of the soul and the combination of the faculties by which is obtained all normal action, and to the fact that the faculty preponderating in quantity in any group gives its character to the group, and that the special character of the associated mental operations is indicated by the form of the group in its composite shape.

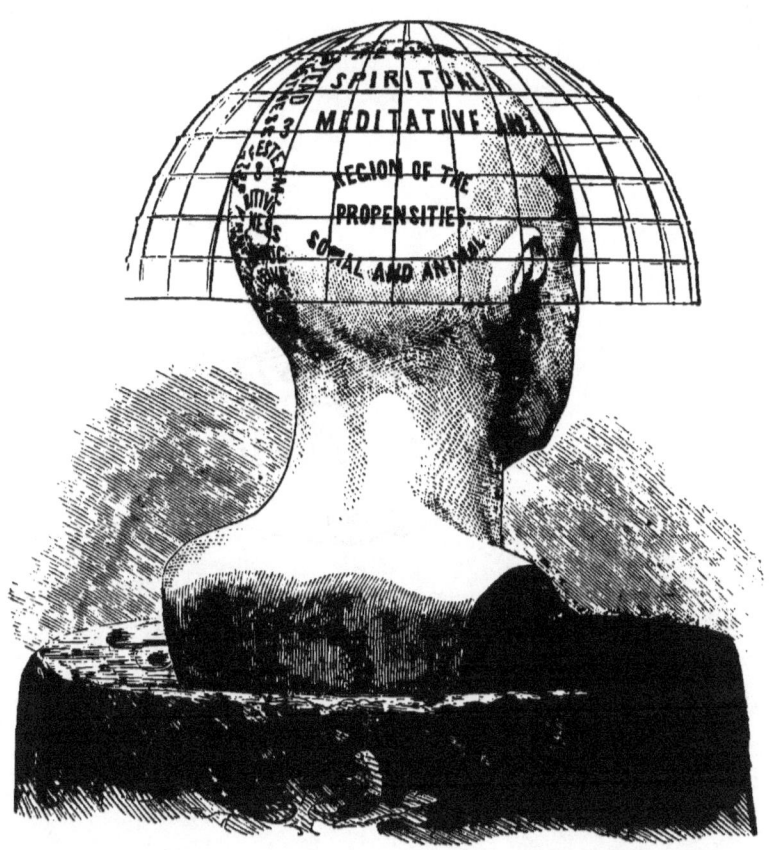

PHOTOGRAPHIC VIEW (CUT IN WOOD) OF THE PHRENOLOGIC BUST OF WASHINGTON.—*No. 2 —Right three-quarters back.*—General localization of the Propensities, and the Restraining Faculties of the Spiritual group. In the lower part of the head the region of the Propensities in either hemisphere is shown. In the upper part of the head, the region of the Spiritual faculties is shown, presenting the Meditative cluster prominent in this view. The lettering which indicates the general regions of the groups is prominent in this view, while the names of the special faculties, which are given in detail upon the other side of the bust, are more fully seen in view No. 5, p. 34.—See pp. 16, 117, 118.

[To face p. 32.]

PHOTOGRAPHIC VIEW (CUT IN WOOD) OF THE PHRENOLOGIC BUST OF WASHINGTON.—*No. 3.*—*Right three-quarters front.*—Localization of the Intellectual Faculties, and the Intuitive cluster of the Spiritual Group. In the region of the forehead and temples the locality of the Perceptive, Conceptive, and Combinative clusters, respectively, which together form the Intellectual group, is shown. In the upper part of the head, the region of the Spiritual faculties is shown, presenting the Intuitive cluster prominently in this view.

The lettering which indicates the general regions of the groups and clusters is more prominent in this view, while the names of the Special faculties, which are given in detail upon the other side of the bust, are more fully seen in view No. 4, p. 33.—See pp. 16, 117, 118.

[To face p. 33.]

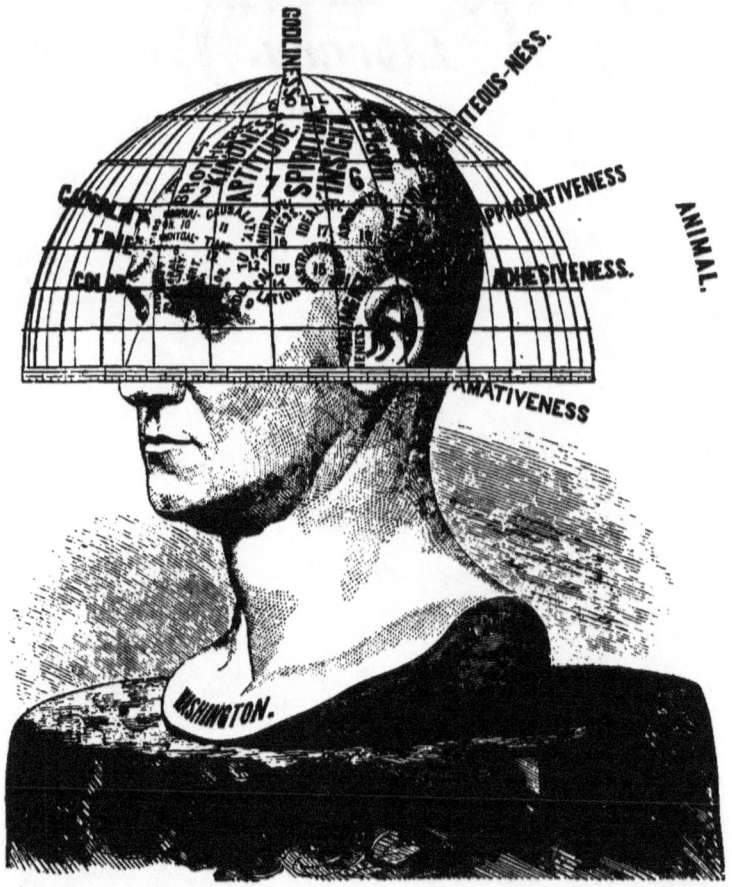

PHOTOGRAPHIC VIEW (CUT IN WOOD) OF THE PHRENOLOGIC BUST OF WASHINGTON.—*No. 4*—*Left three-quarters front.*—Localization of the Faculties of the three groups as exhibited on the left Hemisphere. The special locality of each faculty of the INTELLECTUAL GROUP is seen in the cerebral form manifested in George Washington. In the brow are the Perceptive faculties, viz: Individuality, Language, Form, Size, Weight, Eventuality, Locality, Color, and Order. In the upper part of the forehead the Conceptive faculties, viz; Comparison and Casuality. In the temples and sides of the forehead, the Constructive faculties, viz: Time, Tune, Calculation, Constructiveness, Mirthfulness, Ideality, Acquisitiveness.

The INTUITIVE FACULTIES in the Spiritual group are also seen in this view, viz: Brotherly Kindness, Spiritual Insight, Aptitude, and Hopefulness. The position of the central faculty of Godliness, in the apex of the head, is also indicated.—See pp. 30, 111, 112, 117, 118.

[To face p. 33.]

Classification of the Faculties according to the foregoing Views.

The following are the three groups, and the faculties composing them so far as the functions of the convolutions have yet been ascertained, Except in the first group, the names employed for the faculties are those given by Dr. Spurzheim.

GROUP I.—*The Spiritual Faculties :—Meditative and Intuitive.*

1. Godliness.
2. Brotherly-Kindness.
3. Steadfastness.
4. Righteousness.
5. Hopefulness.
6. Spiritual Insight.
7. Aptitude.

GROUP II.—*The Intellectual Faculties:— Combinative, Conceptive, and Perceptive.*

1. Individuality.
2. Language.
3. Form, (or Configuration).
4. Size.
5. Weight.
6. Eventuality.
7. Locality.
8. Color.
9. Order.
10. Comparison.
11. Causality.
12. Time.
13. Tune.
14. Calculation.
15. Constructiveness.
16. Mirthfulness.
17. Ideality.
18. Acquisitiveness.

GROUP III.—*The Propensities :—Social and Animal.*

1. Alimentiveness.
2. Amativeness.
3. Destructiveness.
4. Philoprogenitiveness.
5. Inhabitiveness.
6. Adhesiveness.
7. Combativeness.
8. Self-Esteem.
9. Secretiveness.
10. Approbativeness.
11. Cautiousness.
* Desire to live.

In giving this enumeration of faculties, I must not be understood as representing it as complete. I am not among those, if any such there are, who believe that Phrenology and Physiology have reached limits beyond which they cannot pass. The investigations upon which a knowledge of *all* the faculties is to be based, are not yet completed. There are convolutions whose functions are yet to be ascertained, which if satisfactorily done, will throw much additional light upon the analysis of the mind, by presenting the organic conditions of some elements of mental character, for which the combinations of the exterior convolutions do not fully account. The results, too, of Dr. Spurzheim's anatomical investigations in respect to the faculties interblending and running into each other, indicates that more extended observations of anatomical structure are necessary, to determine to what extent diversities or variations in this connection of organs exist in different brains. If such diversities exist, they should be investigated before we shall have a full and clear analytic individualization and definition of the faculties that are already known.

Order of dealing with the Faculties in Education. The Restraining Faculties.

In my classification, I place the groups in the order of their importance and due predominance in the matured character, in a Christian and civilized community. In practical education, however, they are to be dealt with, in point of time, in the inverse order. The Propensities must first be brought under control and regulated; and the training of the Intellect, and securing the conditions for awakening the Spiritual disposition, are duties to be successively commenced, by means of the ascendency which the Propensities afford. The order of Education is, to enter the child's mind through the Propensities, (the activity of which depends upon the desires,) thus securing control; by their natural, predetermined force, to develop and regulate the Intellect; and by moral instruction prepare the way for the awakening of the Spiritual faculties, that in the end *they* may predominate, guiding, regulating, and cultivating the Intellect, and holding the desires of the Propensities in check, thus making the whole mind the fit temple of the Holy Ghost. It might appear to some that on this account, education is not required to deal with Spiritual truths. This is not the case, for to the teacher all these truths are of the very highest importance, and the preparation of the minds of children for these truths is the highest object of education.

Since the activities of the Intellectual Faculties depend for their force on the Social and Animal Propensities, it becomes the object of Education to manage these forces, and to direct them to useful purposes. There must be the attention of a teacher, to awaken the faculties of the child to activity.

That the teacher may secure the attention of the child, he should have regard to temperamental affinities, and the physical and perceptive sensibilities. In order to regulate the activity of the faculties, thus awakened, and to fix the attention of the child, and invite or compel perseverance of effort, it is necessary for the teacher to appeal to those faculties in the child's mind which possess a power of restraint over the other faculties, and by which the child's capacity for continuity of mental application is sustained.

The restraining faculties are four in number,—Cautiousness and Secretiveness, which are among the Propensities, and Conscientiousness (or Righteousness), and Firmness (or Steadfastness,) which are in the Spiritual Group. The latter have but little influence upon the mental disposition in childhood, and it is upon Cautiousness and Secretiveness that we must chiefly rely in education, for securing attention, restraint, and the will of the subject; for they are in the group in which the natural forces reside. These restraining faculties, however, when not directed, act against the teacher by selfishness, rather than in his favor. He must first secure their aid, either by obtaining the good will of the pupil by adapting himself to the desires of his Propensities, or by obtaining control of his self interest through fear.

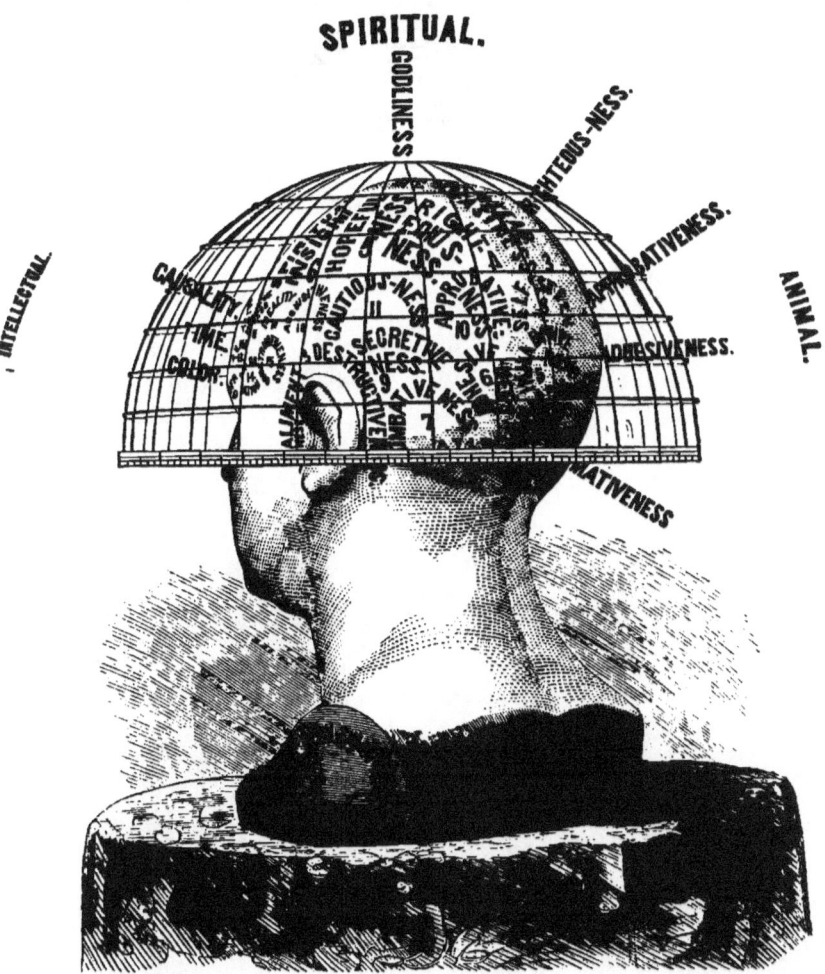

Photographic View (cut in wood) of the Phrenologic Bust of Washington.—*No. 5.—Left three-quarters back.*—Localization of the Propensities, and the Restraining Faculties of the Spiritual group. The special locality of each faculty in the Group of the Propensities is seen here, in the cerebral form manifested in George Washington. They are Alimentiveness, Destructiveness, Inhabitiveness, Combativeness, Secretiveness, Cautiousness, Amativeness, Philoprogenitiveness, Adhesiveness, Self Esteem, Approbativeness. The organ of the Desire to Live is an interior convolution, and its position cannot well be indicated on the exterior of the bust.

The Meditative faculties in the Spiritual group are also seen in this view, viz: Steadfastness, Righteousness, and Hopefulness.

The position of the Restraining faculties in each group is also seen, Cautiousnessness and Secretiveness in the Propensities; Righteousness, and Steadfastness in the Spiritual group.—See pp. 30, 115, 128, 131.

[To face p. 34.]

First impressions upon the Scholar by the Teacher. Sympathy in Discipline.

Hence, while securing restraint through the faculties of Cautiousness and Secretiveness, we must, in order to maintain attention and continuity, and control the will of the subject, rely on Self-Esteem, the Concentrative force, and on Adhesiveness and Approbativeness, as well as on the outward pantomimic activity which appeals to the sensuous life of the child.

First impressions, temperamentally considered, made upon the scholar by the teacher, are of paramount importance as affecting their subsequent intercourse. Every observant teacher will have noticed that different modes of approach must be adopted with different children; whilst an introduction to one at a bounteous breakfast, or over a hearty dinner-table, will secure a preference, a quiet conversation in a retired nook or corner, will tend to secure the permanent affection of another; and whilst the warm regard of one class may be gained by freely mingling in the sports and hilarity of the play ground, a reserved and taciturn air will obtain the confidence and respect of a different class. The restraining faculties of the Propensities are pre-eminently selfish or self-asserting; and to gain ascendency over a child in whom these faculties are predominant, the teacher will have to subordinate them by force or by love. Force, if used, must be sympathetically applied; and if love be used, the teacher must at the same time maintain that authority centralized in himself, which will command respect.

The term "love" in this connection, is used in an enlarged sense, and implies a sympathetic susceptibility on the part of the teacher, in the suffering the child endures from the infliction of punishment; and this sympathy must be so clearly manifested that the child cannot fail to perceive and acknowledge it. The influence of such an exhibition of feeling must be attended with the happiest results. For, even though tender and immature, the sensuous mind of the child, delicately influenced, as it is, by vivacious outward expressions of feeling, either of pleasure or pain, will appreciate this manifestation of suffering on its behalf, and when it sees that the infliction of punishment is painful and distressing to the teacher, on its account, and does not arise from a selfish desire to gratify angry passions, a sense of grateful obligation will be awakened, rather than emotions of revengeful dislike, and a determination will be formed not to wound again one who sympathizes so acutely in its own sufferings.

And as this regulated sympathetic sensibility can exist only as a moral, persuasive influence, and then only to a limited and partial extent, in an unconverted teacher, how important it is, that the power of the Holy Spirit should be shed abroad in the hearts of those to whom the education of the young is entrusted.

It is necessary to observe, however, that the love which arises in the child's mind is that of the selfish and social faculties, which are grouped together in the Propensities. Appeals to moral grounds, in the first instance, fail to have any direct bearing on the child, until the teacher has secured its confidence, respect, and docility, which can be done through the Propensities by the aid of the Intellectual faculties. Through the faculties of self interest and social happiness, which are in the group of the Propensities, the teacher may acquire, as far as a stranger can, the temperamental as well as the mental influence possessed by a parent, the origin of whose power is in the natural, instinctive sensuous activity of the temperaments and these sensuous faculties. To this end a wholesome fear is necessary, both for immediate control, and for the ultimate development of a Godly fear, which will depend upon the awakening of the spiritual disposition.

In the choice of the motives by which the teacher will endeavor to attain this control, he should, in general, give preference to the upper range of the Propensities,—*viz:* Cautiousness, Approbativeness, and Self-Esteem,— which lie contiguous to the Spiritual Faculties of Righteousness and Steadfastness. Cautiousness can generally be best awakened in the pupil by appealing to the next contiguous faculty in the same range,—*viz:* Approbativeness, or if this is not the next largest, to Self-Esteem. If the teacher allows the lower range to be most actively exercised, he increases the difficulty of training the Intellect; for Secretiveness tends to make the child selfishly watchful and sly, and Combativeness is the rebellious spirit. The natural tendency to active development in boys is in this direction. In girls the predetermined tendency to development is in Philoprogenitiveness and Adhesiveness and in the upper range, *viz:* Approbativeness, Cautiousness, and Self-Esteem. It should not, however, be forgotten, that the fear awakened in a child's mind, although necessary, is a selfish fear, and not to be trusted to control the child when it is not conscious of the teacher's cognizance. Fear, or Cautiousness, is but one faculty; and it will be understood that the largest quantity of brain matter, or number of faculties or motives, that a teacher has in his favor, the easier it is to command.

Some children, however, from inherited organic form, insufficient means of growth, or a sentimental and unpractical home training, will be found deficient in this lower range of Propensities. In such cases, the teacher should call them into exercise. The boy may be encouraged to destructive sports, such as fishing and gunning, to kill his own food, thus developing Destructiveness, Alimentiveness, and Combativeness, so that he may have greater executive force of character. The more humane way to attain the same result, is to put the boy into the society of boys in whom those faculties are large, by whom he will be compelled to defend himself and exercise executive power, and to encourage him to take his own part. This training tends to give greater practical vitality, energy, and versatility; but, except where it is necessary for correcting absolute deficiency, this gain may be at the expense of intellectual talents and attainments.

By a knowledge of these sensuous faculties, which will be more fully discussed in answer to the later questions, the teacher may intelligently gain possession of the will of the child, whether it be centralized in the socia

Faculties may be considered Individually; but their action is Associate.

faculties, leading the child to be tractable, or whether it be centralized in the personal faculties, leading the child to be self-asserting. He will use for that purpose such faculties as the child's organization and needs of development indicate to be most proper; and by means of this ascendency will be enabled to train the faculties in the Intellectual group, developing their powers, according to a proper order, and preparing the way for the maturity of the whole being.

CHARACTERISTIC ACTION OF THE FACULTIES.

The activity of a human faculty may be theoretically considered either as *individual*, or as *associated*. It might be in fact *individual*, if it could spring from and express, merely the vigor of the faculty itself, uninfluenced by any other faculty, either of the same, or of a different group.

But, as we have seen, each faculty is *associated* with the surrounding faculties contiguous to it, and is more individualized, only in proportion as surrounding faculties are less developed. It is modified by the action, whether restraining or non-restraining of another faculty of the same being and in the same group, and quite changed or compounded in character by a faculty in either of the other two groups.

Although the activity of the mind may be centralized in one faculty, so that the qualitative character of that faculty predominates in the mental act, mental activity is not engrossed in one faculty to the exclusion of all others. The *external* act resulting from an activity of the mind is dependent upon the executive influence of the faculty of Destructiveness. The *internal* or mental act is the peculiar condition or composite activity of the whole mind, apparent in the organic form of the brain, and in the pantomimic phases of expression.

An instance of individual activity, in actual life, would be monomania. What is meant by individual activity is, that in analyzing the mental processes, we are to have regard, first, to the activity of each faculty, considered by itself, bearing in mind its size, the group to which it belongs, and the group in which the largest faculties may be. But the observer must keep in view, secondly, that the force, activity, and restraint, are naturally centralized in the Propensities, with modification by temperamental disposition; thirdly, that all mental action is composite, and modifications of activity are produced by associated faculties, in their combination; fourthly, that further modifications are produced by the sympathetic action of other minds; while lastly, the most important modifications are produced only by the Grace of God in the soul, manifested through the Spiritual group, giving, by the meditative part of that group, illumination and prevenient knowledge,—prophecy.

Now the organs of these mental processes, as they are actually presented to us, we perceive to be associated in three groups,— those of a congenial nature being together;— and the faculties belonging to each group, we perceive to be usually in action simultaneously, one or the other predominating, while its associates co-operate with, and stimulate or restrain its action. The mental act thus exhibited is the result of the compound action of the faculties so associated, (whether they belong to the same or either of the different groups), not the independent act of one.

In our analysis, the character and activity of any faculty, individually considered, must be regarded as chiefly depending upon the physical contour, shape, and structure of the brain organs. The peculiar form of the organ must receive careful attention, both as to height, breadth, and shape, and the direction of development from the given centre of the head in a line passing through the ears. If either side of the organ predominates, the organ is modified upon that side, by the influence of the organ next to it. This constitutes the endless variety in each organ, and gives rise to the variety of manifestation.

Not only, however, the structure and size of the organ, but also the temperamental tone, the period of life, the health, the supply of natural vital force, the degree of imparted grace,—all these qualities taken together, form the conditions on which the individual activity of each faculty depends. As the explosion of powder depends for its effect, not alone on the quantity of the powder, or its fineness of quality, nor on the size of the grain, or the degree of compression, or the shape of the instrument, but upon all these things in combination, so also the energy of a faculty depends upon the physical qualities of the organs through which it acts, and the temperamental conditions of the bodily organization, in accordance with the general law to which I have made reference, that temperamental conditions being equal, the size of the brain is only a measure of its power, its activity is the measure of its influence.

Our knowledge of the associate action of the faculties depends largely on the observation of the relative size of the contiguous faculties in their special order in groups; the facility with which they interchange and continue effects upon each other depends on the general structure of the brain and the special structure of the faculties, on their relative vigor as compared with each other, and as measured by that of the organs, and on the influence of the temperaments.

Hence, in our examination of the individual activity of the faculties, we must observe which is the largest group, (that giving general character to the mind), and which is the predominant faculty in the group, (this giving mental character in the group in which it appears), and we must also consider the physical peculiarities of the organic structure of the brain, and the phases of quality of mental action arising from the combination of temperaments in the person in question, whether more acute, by reason of predominance of the Nervous, or more warm from the Sanguine, or more placid from the Lymphatic, or more firm and metallic from the Bilious; and the result should be considered with reference to the age and previous culture of the individual.

Age and period of life, in which the temperaments of the body alternate, have a marked influence. In infancy, the activity of the Propensities predominates with the Lymphatic temperament, in developing the body in growth, and this temperament, when intervening between the Nervous and Bilious, or between the Sanguine, if that be predominant, and either of the others, can be increased or diminished, when properly understood, so that inheritable predispositions may be suitably regulated by parents or the teacher, before the period of puberty. From puberty to manhood, or middle age, the Intellectual life becomes settled or mature; from middle life to old age, the Spiritual Faculties tend to take precedence, by the declining activity of the Propensities; and each of these epochs has an important general influence in directing the faculties.

In this realm of inquiry into the individual and associated faculties, we have not referred to the Spiritual life implanted by the Holy Ghost, but have been mainly guided by the physical peculiarities of the predominating organs of the brain,—whether considered individually or as co-operating in one and the same composite act,—and by the physical peculiarities of the temperaments.

Although we thus deal with the organs as indications of the qualitative character and of the order or predominance of the faculties, we are not to regard the faculties as the products or results of the organs. The organs are the conditions but not the efficient causes of the mental functions. Dr. Gall defines an organ to be " the material condition which renders possible the exercise or the manifestation of a faculty." And he adds, " According to this definition, it may be conceived that no exercise of a faculty is possible without an organ, but that the organ may exist without the faculty to which it belongs, being put into exercise."

We recognize in the soul, an existence independent of, and superior to the brain organ. The physical condition, which the organ constitutes, limits the manifestation of the corresponding faculty only in respect to its qualitative character. There is also a predisposition to activity in special faculties arising from inherited size and structural order of the organs, and from the fact that the nerves of sense have immediate communication with the organs of the sensuous faculties, the Propensities and Intellect, through which the mind is made subject to sensuous influences. This subjection, which does not result from defective organization, but from engrossment of the mind in external and sensuous objects, is the fallen condition of man. Hence the necessity of Divine influence, to awaken the soul to spiritual life, in the faculties which are manifested through the organs of the Spiritual group, which have not the same direct sensuous communications. God acts directly upon the soul, giving light, by the faculty of Godliness, through the Spiritual group of faculties. Man's actions proceed from his Propensities in combination with the Intellect and the moral sentiments.

The impressible and sympathetic activity of the faculties must, therefore, next be considered; and in this realm, physical laws no longer guide us, but we must have recourse to spiritual ones. For, so far as the activity of the faculties is sympathetic, it depends on the relations established between the soul and other beings, and on the power and direction in which those influences of others are acting, which stimulate, check, or otherwise act upon the faculties of the individual, and these, in their higher and proper order, follow spiritual laws. When the Holy Ghost thus influences a mind sympathetically disposed, the influence is direct and all-absorbing, and may be said to change the nature of the faculties and the whole being. Neither class of laws, however, can be adequately understood without reference to the other. Their operation is inseparably interdependent in the human life. The peculiar phases of the existing combination of temperaments, as well as the peculiar order of cerebral organization, qualify the external manifestation of the sympathetic and moral qualities.

Up to the age of seven, more or less, varying in accordance with the law of inheritance, the work of education naturally depends upon the MOTHER for its sympathetic continuance. At the period when the mother's responsibility diminishes, the TEACHER takes it up, and should commence and continue it *under the same conditions of delicate physical sensibility, and temperamental sympathy*, which result from the characteristic lacteal lymphatic temperament of woman.

In our fallen nature, the Social and Animal Propensities, with only a qualifying influence from higher sentiments, ordinarily possess control, both of the teacher and the scholars; and the conflicts arising from the predominance of these self-asserting faculties, make the vocation of teaching most difficult and wearisome. And where the Spiritual faculties are predominant in the teacher, ignorance of the temperaments and of the science of the mind will lead him to neglect the power he possesses in this respect, or render him unable to apply it with the sympathy in sensuous conditions, which is necessary for securing control of the Propensities of the child.

Hence the influence of the religious father or teacher is so commonly irksome to the child; while the sympathies of the mother, which are divinely ordered in sensuous attention and in the predominance of the faculties of Philoprogenitiveness and the other Social Propensities, are congenial to him.

When Christianity is properly understood by parents and teachers, and the influence and guidance of the Holy Spirit is humbly sought in intelligent conformity to the laws of the mind, the Spiritual, Social, and sensuous sympathies may be made unitedly and harmoniously active. This is the great and immediate necessity of Society in respect to Education.

A limited and partial success, in teaching, may be attained by natural gifts; but however great the natural gifts of the teacher, and however earnest his efforts, his success, especially in the higher departments of moral and of spiritual training, will be limited and partial only, unless, as a means of acquiring power, he is under relations of sympathetic influence by the Holy Spirit; and unless, as a means of exerting it, he is under similar relations with those whom he teaches. The true teacher is born of the Spirit, for his vocation.

These relations of sympathetic influence are established through spiritual impressibility and susceptibility. In fact, as a means of education, these qualities are of paramount importance in the teacher. A child with moderate Firmness (or Steadfastness,) and Conscientiousness (or Righteousness,) if influenced by a teacher in whom these spiritual restraining faculties are predominant, will do better than a child in whom these faculties are large, under the influence of a teacher in whom they are passive.

In No. 1 of the *Unit*, you will find a delineation of the character of a gentleman, which furnishes a good illustration of the defects of any system of education which is not founded on the activity of the Spiritual Faculties. If they do not lead in the work of the teacher, under the guidance of the Holy Spirit,—if he does not find in the impressive and sympathetic influence of the Holy Spirit of God, the source of his power, and the stimulus of his activity, he is of necessity left to depend upon the Propensities as the impelling force in his character. What is referred to, in the character there delineated, as *boyishness*, is simply the predominant development of the Propensities, social and animal, under the Lymphatic and Sanguine temperaments, in the order of combination characteristic of the organization manifested in boys,—Alimentiveness, Destructiveness, Secretiveness, and Combativeness;—and with the sensuous quality given by large development of the nerves in immediate connection with the brain. These faculties are earlier developed than the Intellectual and Spiritual groups, and are, in themselves all-important to a manly character; but whenever they maintain a predominant influence after the age of puberty, overruling the moral qualities when the latter ought to acquire sway, the character retains this phase of boyishness, although favorable circumstances, social influences, temperamental character, etc., may restrain the individual from the evil courses to which they naturally tend.

This sensuousness gives immediate force, energy, and momentum to the character, but exposes its possessor to great temptations; and this sensuous condition, when continued beyond the period of childhood, and without the Spiritual disposition, is the seat of all that is vicious, depraved, and criminal in human life. And thus, the result of a system which ignores spiritual influence as a source of strength to teachers is, that some of our most efficient and successful teachers must come from among those who are most strongly predisposed to temptations.

III. *"Would Not Education, then, as a Training and Developing Process, be Based in its Practical Operations, upon—1st. Peculiarities of Temperament; 2nd. Peculiarities of Cerebral Structure?"*

I answer: that inevitably, and not by the intelligence of the teacher, it is already so based, in many of its processes and operations; and the first requisite to more successful teaching is an intelligent attention to the peculiarities of temperament and of cerebral structure, and a scientific adaptation of the processes of instruction to those physical peculiarities.

But education cannot be wholly based upon these physical conditions, We must recognize the fact that although physical organization is the condition, it is not the efficient cause of mental activity. Hence the processes of education must be based also upon an intelligent consideration of the stimulus and influence received by sympathy with other beings, through the Social, Intellectual or the Spiritual group of faculties. Of these external influences acting upon the child, the teacher should take care to be himself the chief, and this can only be fully and successfully done by a knowledge of the Spiritual laws to which I have referred.

A survey of human nature, in the aspect both of its physical organization and its Spiritual life, shows the chief progressive stages of human education, viewed in its largest and fullest sense, to be chiefly these:

First. Such a culture of the physical organization as will carry the individual activity of the faculties, under favorable temperamental conditions, to the highest perfection consistent with their true proportional or harmonized action. This object is promoted, to speak in general terms, by affording the enjoyment of nutritious and healthful food, and suitable and varied exercise, both of body and mind,—by all those influences, in fact, which tend toward the vigorous action of the digestive system, the thorough æration and vigorous circulation of the arterial blood, and sufficient bodily repose for the liver and the equalizing functions which it maintains,—thus supplying the proper harmonized conditions necessary for the most healthy growth and action of the brain. This is the domain of *Physiology.*

Second. The proportional or harmonized development of the faculties, or rather of their brain organs, considered with due regard to their combinations in groups, clusters, and special associations. This object is promoted by all those methods which regulate the exercise of the mind according to the order of the faculties, as numerically marked on the bust, and to their relations in groups, always noticing that the mind must be approached by and through the predominating faculty of the group, so as to get the attention of the person taught. In this way, we may enlarge and strengthen those organs which by nature are too small, and diminish in size or activity those which are naturally too large, thus establishing an equalized fulness of mental operation, and developing the faculties (in their appropriate groups and clusters,) in the proper order as delineated on the bust. This process approaches the mind first through the sensuous nature, and according to the various functions of special sense, *viz:* sight, smell, taste, hearing and touch. To give the most vigorous force to this process, children are trained together, thus bringing into play the social or family feelings,—*viz:* Adhesiveness, Philoprogenitiveness, and Inhabitiveness, and, subordinately to them, the power of Emulation, *viz:* Approbativeness. This is the peculiar business of "*Teaching*," or "*Education*" in the restricted sense in which the latter term is popularly used.

Third. The establishment and maintenance of such relations and social conditions among men, as that the individual shall live, as much as possible, under conditions of sympathy with those of his fellow men whose faculties are acting in the order of their proper development, and who are in a condition of susceptibility to the influence of the Holy Spirit; and that he should not be influenced by sympathy with those whose sensuous and sinful habits would lead him astray. This object is promoted by all those methods which draw men together in the common and sympathetic prosecution of a worthy purpose; by the organization of meetings, and of institutions for co-operative action; by customs and usages which excite emulation and ambition; and by the sequestration of the criminal and vicious from the general society of men. This is within the scope of *Social Science.*

Fourth. The establishment of such relations between the soul of the individual and the Holy Ghost, that the direct influence of the divine Spirit, operating through Spiritual love, upon the human soul, may be recognized by man, and may be consciously and unhesitatingly accepted by him as the guide of his actions. In the attainment of this state, two things are needed,—a susceptibility to the Divine influence, and a willingness to be influenced by it, and by those who are themselves under the same influence. All moral agents possess a degree of spiritual impressibility: to be wholly without it, is to be not in the category of beings morally accountable. But it differs in degree, according to the activity of the Propensities, and according to the culture the Intellect has received. Moreover one individual may be highly susceptible to the Divine influence, yet, by his natural and inherited perverse will, may wholly resist it. Another may be cordially willing to yield to it when recognized, yet able to feel it but feebly, by reason of long indulgence of the Propensities and Intellect. A third may feel it fully, and yield to it cordially. The willingness to be thus influenced comes through the inward consciousness of good and evil.

To change the natural unwillingness of man to yield sympathetically to the influence of the Holy Ghost, into a cordial and earnest submission to it, is a work in which other men become merely instruments. It lies between the individual himself and the Holy Spirit of God. The conditions under which this change must take place are as the wind which "bloweth where it listeth." He gives Spiritual consciousness alike to all who will receive life. But this change must be promoted by human agency; and to do this—to bring to the knowledge of men the laws and methods by which God acts upon the soul, to acquaint them with the motives and reasons which should induce them to submit to His influence, and to encourage them in the effort to yield their faculties to His dominant control, is the office of *Religion*, and is the task especially committed to the Church of Christ. The science of the mind is the handmaid of true Religion. The present intellectual state of the world makes this objective knowledge of the mind of fundamental importance to Religion at this time.

Education, as a training and developing process, should be based, in its practical operations, not alone upon the peculiarities of temperament and cerebral structure, but also upon the application of the foregoing principles, which are fundamental to the ultimate welfare of both teacher and pupil.

These general principles are the foundations of the answers which are to be given to the remaining questions in your letter. I will, in a short time, attempt the answer of the remaining questions in detail.

Very respectfully,
JOHN HECKER.

Description of the Colored Illustrations of the Temperaments.

[*From Mr. Hecker to Mr. Kiddle.*]

HENRY KIDDLE, Esq.,

DEAR SIR:—In a former letter, I have answered the first three questions proposed in your letter of July 27th, and have shown that, inasmuch as there is a spiritual as well as a physical nature in man, the processes of education must correspond with this dual nature; there must be a recognition of the physical part, and a treatment of it by physiological methods, and there must likewise be a recognition of the spiritual part, and a subjection of that to spiritual influences. Even at this intelligent period of the world, it is the error of all educators to confound these two natures.

I now proceed to answer the questions embraced in the second division of your letter, marked (A.)

All the questions in this division relate to the temperamental conditions of the body, and to the modifications which a just attention to the temperaments introduces into the means and processes of education. It is difficult to describe in language only, that which should be demonstrated practically; but I will make such explanations respecting them as I am able to do in writing.

In my last letter I described a plaster bust which I have had made. to show the phrenologic development to a better advantage than has heretofore been done. In a similar method I have prepared illustrations to show the temperamental peculiarities. They consist of four portraits of George Washington, each printed in oil colors, after designs in water colors, from photographs, according to the characteristics of one of the four temperaments, so far as those characteristics are manifested in the expression of the head. These will illustrate and individualize the peculiarities of each temperament, so that in studying them, the mind may be enabled to discriminate them more perfectly than merely verbal description suffices to do. Similar exemplifications of the whole structure of the human body would be still more useful, but are not at present practicable. I hope that some day it may be in our power to have them. The manikins which the French have produced, for teaching anatomical structure, would go far to serve this purpose. The account of the Physiological functions given by Dr. Draper, in the work referred to above, and also in his smaller text book, gives the interior conditions of temperaments, without, however, any systematic statement of their external phases.

If I were merely describing the temperaments I should take four different figures for these illustrations, for there are relations between each temperament and the special form of the head; but as it is essential to a knowledge of the mind to discriminate between the modifications of disposition dependent upon the temperament, and those dependent on cerebral structure, it has seemed better, first to delineate the various temperamental characteristics, and to take up cerebral differences afterward. The reader must, however, remember throughout, that the contrasts of color presented by the illustrations would in actual life be inseparably connected with some contrasts of cerebral form.

Asking you therefore, to refer to these illustrations, from time to time in the course of the description, for assistance in the elucidation of the subject, I now proceed to answer the questions propounded under division (A), in your letter.

"1. *Would the Division of Temperaments into the Four Primary Classes be Sufficiently Minute, as a Basis, Without Taking into Consideration the Various Combinations as they Occur?*"

In the proper sense of the term, the "*temperament*" is the precise mixture or combination of physical qualities belonging to each individual. This presents so vast a variety, that if we should go into detail without limit, we should lose ourselves.

For a Scientific account of the temperament, it is necessary, as Mr. Bain suggests, "to describe as well as can be ascertained, the peculiar condition of every one of these (temperamental) organs *seriatim*, drawing the proper inference, without inquiring which of the four temperaments the case falls under."

But to do this would be beyond my present limits, as well as impracticable in popular application; and I must, therefore, use the four commonly understood names, representing characteristic extremes of physical constitution, for the purpose of clearness. There is too much discrepancy without proper classification, in the descriptions given by Physiologists, to enable the general reader to enter into a more precise delineation.

It is desirable, and will tend to a still further improvement of educational science, to take the intermediate phases into consideration; but in initiating the effort, too much must not be undertaken at once, and what is undertaken must be clearly and objectively defined. It is best first, to individualize and clearly define what are known as the four primary temperaments, and defer the combinations until a later stage of the effort.

It is certainly within my hope and expectation that the time will come when the ordinary combinations of the temperaments will be well understood, and their physiological indications familiarly known to observers; and when methods of instruction and discipline will be adapted to many different varieties and shades of combination. The principles I advocate cover the whole ground; and I stand ready to show the application of them to minute classifications and subdivisions of temperamental character, to as great an extent as it is found that the teachers become suf-

What is meant by "Combined Temperament."

ficiently accustomed to the requisite observations, and possess the adaptability to apply them. But I have at present only urged a classification of pupils by the four original temperaments, from an impression that this is all that can practically be accomplished at the outset.

In speaking of these original temperaments, therefore, it must not be forgotten that in truth, all persons exhibit a combination of temperaments. An instance of a pure and uncombined temperament, if not indeed an impossibility, would be out of the common course of nature. To conceive the entire exclusion of either of the primary temperaments, speaking in the strictest sense, would involve the supposition that the vital organ which imparts it, the brain, the lungs, the stomach, or the liver, as the case may be, were wanting in the system. Life, in the human organization, is made up of the four temperaments, and requires some admixture of the qualities of all of them. When we speak of a person as exhibiting a given temperament, we mean that his constitution departs from the harmony of the equal order, by an excess of one temperament predominating so strongly over the others, as to rule and lead throughout the whole constitution. The predominating temperament marks its influence in all the others, imparting to them something of its own quality, either sharpness, or warmth, or placidity, or cool, staid force, as the case may be. When we speak of a combination of two or more temperaments, we mean that all the four so compare in force and development, as that those two or more designated in the combination, in the same way exercise an individualized as well as a blended influence.

The physiologic processes of the Nervous temperament distinguish animal from vegetable life. Each of the three other temperaments, the Sanguine, Lymphatic, and Bilious are characterized by chemical and mechanical processes similar to those of vegetative life. Their functions are unconscious, but by their union with the nervous system they are all brought into a general relation of sensibility in the mental consciousness residing in the brain.

In this union the function of each temperament acts on and is re-acted on by the others, and the mind, through the nervous system, is divinely ordained to manage and direct the whole.

Diseased conditions of the body change the temperamental relations, and changes in temperamental character may be effected by the administration of appropriate medicines; indeed, the whole system of the Materia Medica, when properly understood, clearly demonstrates this proposition.

In the fallen state of man he is given up to the dominion of the three lower temperaments and the sensuous phases of life, engrossing the nervous forces in the unconscious functions of the temperaments, and centralizing the mind in the Intellectual faculties and Propensities which lie at the base of the brain, in sensuous communication with the body and the external world. Hence, the necessity of the intervention of a higher power, the Spirit of God, to regenerate the nature, and give spiritual light that shall intelligently and consciously direct the life through the Spiritual Faculties, in accordance with the law established in God's original creation of man.

It is always to be borne in mind, therefore, that the influence of the Holy Spirit, which is given to man through the Spiritual part of his dual nature, modulates the activity of the temperaments, and harmonizes their influence in the combination. The brain and temperaments present the physical conditions of human life; but it is only when the Holy Spirit awakens the Spiritual Faculties, that man possesses the power of his true and full life; and by this influence only, directing the mind, is given the power to overcome the inertia of matter, and the body is made illuminated and refined, transcending any physical manifestation.

In order to show that it is practicable for ordinary observers to distinguish the four temperaments, I will now mention the leading physiological peculiarities by which they are marked.

In each temperament some one function or system of organs, characteristic of that temperament, leads in the organization, influencing all the others, and producing modifications of color, shape, size, and texture. This general principle was alluded to in a previous communication.

One who examines the subject for the first time will find in his observation cases that apparently form exceptions to any general description. Thorough acquaintance with the causes of temperamental character, in combination with cerebral causes, will explain every exceptional case.

In my descriptions, whether of size, form, color, or activity, I must be understood as referring to the white races and the healthful state.

THE NERVOUS TEMPERAMENT.

By the Nervous temperament, we do not mean that weakness or excessive irritability, which is commonly called Nervousness. The Nervous system—the direct instrument of the brain, is that which distinguishes the mental processes of the brain from the sensuous processes of the temperaments, and also marks the distinctive characteristics of the sensuous operations of the other three temperaments, the Sanguine, Lymphatic, Bilious—as compared with those of brain.

In the Nervous temperament, the brain, which is the organ and centre of mental life, together with the system of sympathetic, sensory, and motor nerves associated with it, maintaining communication with the organic and temperamental functions of the body takes precedence over these other functions of the system, in respect both to proportionate size and to activity.

Physiologists who have pursued their investigations of the system of nerves acting in conjunction with the brain, without regard to the truths elucidated by Drs. Gall and Spurzheim, have not sufficiently defined the

NERVOUS.

Organic condition of the Nervous Temperament.

sympathetic and sensuous communication existing between the body, with its other three temperaments, and the brain, through its sustaining and commanding influences. They have not discriminatingly distinguished the influence of the mind upon the body, nor sufficiently regarded the fact, that the functions of the Nervous system are centralized in the brain, or are dependent on or controllable by the brain, which is the seat of sympathetic influence, restraining power, and motive and impelling force.

The distributed influence of the Nervous system throughout the surface of the body, and the localized action of the organs of special sense, have each been abstractly considered as forming a system by itself, without due regard to the controlling forces resident in the brain.

In the Nervous temperament, the head is relatively large, and the thoracic and abdominal regions small. The whole nervous system, including the brain, being predominantly active, the mental manifestations are proportionately active. The sharp, keen, nervous sensibility which is characteristic of this temperament,—the brain being the organ and centre of all mental life,—is not owing to the nerves being deranged, or delicate, or weak; on the contrary, their action is disproportionately powerful; for the nervous system, predominating in the organization as compared with the other three temperaments, its action is not sufficiently qualified by the influence of the Sanguine, Lymphatic, or Bilious temperament. Each of the temperamental systems has, within itself, its own appropriate qualifying and modifying and accelerating influences; but the Nervous temperament is superior to the others in that its restraining forces are organic, consisting of the faculties of Secretiveness and Cautiousness, Steadfastness, and Righteousness or Firmness, possessing within itself the will-power of all the other temperaments, when properly understood and applied; while the qualifying and modifying forces of the other temperaments are functional. Nervousness, as usually noticeable by others, is not a preponderance of the Nervous temperament. Nervousness, when it is not caused by ill health, that is, by a derangement of either of the temperamental systems, arises from excessive activity of the senses and the weakness of the restraining faculties in the brain. These faculties may and ought to have full development and influence where the Nervous temperament exceeds the other temperaments. This is the case with all men of the greatest intellectual influence in society, and this is the condition of their power.

In persons of the Nervous temperament, the organs of the brain being larger in proportion, and functionally more active, than the lungs, stomach, or liver, and the bones and muscles being relatively small and delicate, the effect of the mind upon them is greater than in persons in whom the other temperaments have the ascendency.

If there is a great predominance of the Nervous temperament, it is apparent that the life of the person is chiefly in the brain. If also the restraining faculties are large, this gives great self-serving power, and those

External Indications of the Nervous Temperament.

qualities which enable a man to lead and control others. It is this temperamental disposition, thus centralized in the brain, which possesses that commanding, inspiring, and controlling personal presence that compels and moulds the wills of other men,—a quality commonly spoken of as indefinable and undescribable, but which consists in this peculiar combination of mental faculties with the Nervous temperament.

In persons in whom the Nervous temperament predominates, the hair is fine and silky, and brilliant in expression. In childhood it often tends to a transparent whiteness, and, though it grows darker about the age of puberty, it usually remains light in color as compared with that of persons of the other temperaments. It is not abundant; on the contrary, it is often thin and sparse, and has a tendency to fall off early. The skin is thin and transparent. The eyes are bright, vivid, and expressive; quick in their movements, and brilliant and deep in color, usually tending towards gray.

There is a general fineness of quality characterizing the whole physical structure. The figure is delicate, and there is a tendency toward slenderness, and a transparent expression of the whole bodily conformation. The features about the head, particularly the chin and nose, are sharp, well-marked, yet delicate, and the brain development is clearly defined. The muscles are small, but well formed, and indicate an active condition, and their movements are marked by rapidity and promptness. This slenderness and leanness often amount to positive emaciation, and give an appearance of delicate health, while the real condition of the body may show this appearance to be erroneous, where mental pursuits are the vocation. This is shown by the longevity and power of persons of this habit of body, when the Spiritual faculties are in proper command of the Nervous system, and are truly regulated by their Spiritual and sympathetic nature.

In estimating the force of the Nervous temperament, we must regard the cerebral organization, the ganglia at the base of the brain, and the size and form of the neck, which contains the connection between the brain and the nerves of the body. The mental and physical systems reciprocate through the ganglia at the base of the brain.

In a person of the Nervous temperament, if these ganglia, and the faculties at the base of the brain are large, they give a sensuous character to the action of the mind, and indirectly impart vitality and endurance to the whole of the nervous system. This has been called by some the vital temperament. Under these conditions, the brain, the organ of mental life, imparts to the nervous system a peculiar vividness in all the senses, quickness and sharpness of muscular exertion, and a fineness in the details of outline and feature.

Yet, without the spiritual power given to the mind by the influence of the Holy Ghost, the more the brain predominates in the organization as compared with the lungs, the stomach, and the liver, which are the seat of bodily or physiologic life, the sooner its power is expended by action, causing exhaustion, and requiring cessation of activity for recuperation.

SANGUINE.

THE SANGUINE TEMPERAMENT.

In the Sanguine temperament, the lungs and arterial system, which are the organs of atmospheric life, expanding the blood with oxygen, are predominant. In this system is the seat of the mechanical force which compels the circulation of the blood; and, if the organs of circulation have a proper physiological structure and size, they will compel a vigorous circulation, the pulse will be strong, and all the external indications of the arterial system strongly marked. In persons in whom the Sanguine temperament predominates, the thoracic region is relatively large, the lungs being large and the respiratory muscles well developed.

The blood, in its aerated condition, being expanded, active, and diffusive, interpenetrates, nourishes, warms, and stimulates the nervous, lymphatic, and biliary systems, making the muscles round and well filled; and the organs of motion being thus stimulated by the vitalized blood, muscular exercise is enjoyable, and industry natural. The brain, which is not alternating or intermittent in its own power, partakes of the influence of this general pulsative, or, as it were, spasmodic state; and the activities of the stomach and liver are qualified by an infusion of the genial, pulsative warmth of this Sanguine temperament. The organization is characterized by a refined vigor, and a facility in its functions. There is a quick and volatile activity and expression in all the senses and in the movements of the body.

In persons of this temperament, the hair is red, the eyes are blue, the complexion is ruddy, and the skin throughout is fair, and suffused with a reddish tinge which shows the aerated blood to be actively and abundantly diffused. It is this activity or fullness of the circulation wh ch gives brilliant red to the complexion, sometimes imparting a fiery expression. The cheek flushes quickly and readily with exercise, or the varying emotional actions of the mind. The face inclines to roundness. The countenance is animated. The chest is full, high, and expanded. The limbs are plump. but tapering and delicate, with hands and feet relatively small.

The size and vigor of the lungs are, however, the leading indications of this temperament; for it is possible that an individual may have, by inheritance, the chest and muscles of the chest large, and a ruddiness in the skin and hair, and thus present many superficial indications of the Sanguine temperament, and yet the lungs be, in fact, without special energy, so that the individual has not the peculiar warmth and animal vigor which the Sanguine temperament presents. The force of the circulation, which, as has been stated, is the principle of the Sanguine temperament, so far as it is dependent on the lungs, is governed by their mechanical and physiological structure, and their susceptibility to change of blood by atmospheric influence, rather than by the dimensions of the chest.

Persons of the Sanguine temperament are easily influenced by immediate causes, and are volatile in character.

THE LYMPHATIC TEMPERAMENT.

In the Lymphatic temperament, the stomach, which is the leading organ of the digestive system, and the Lymphatics, which gather and convey the lymph liquid from all parts of the system, are predominant.

In this temperament, the stomach and intestinal canal are generally large. Through this channel are introduced the chief supplies of material for the body, and it constitutes the leading physical function upon which the liver depends for its supplies for the nutrition of the whole system.

The fluid of the Lymphatics consists partly of the unused plasma or watery serum of the blood, and partly of the products of waste in the tissues. Into this stream of lymph, brought from various parts of the body, the lacteals, which are connected with the Lymphatic system, bring the chyle or fat digested from the food, and the mingled fluid is poured into the veins near the entrance to the heart.

If this temperament is predominant in the organization of the person as compared with the other three, there is an undue preponderance of the glandular system; and the current or flow of the circulating fluids, though abundant, is generally sluggish. The abdominal region, including the stomach and intestines, is large, and the brain and thorax relatively small. In the excess of this temperament, lymph liquid and chyle exist more abundantly than the functions and reciprocal relations of the liver, the lungs, and the brain require; hence results sluggishness of nature. The cellular tissues are filled to repletion with the super-abundance of liquids. The muscles are burdened with a useless load, which renders their action slow and difficult. The brain is retarded in its action by the same influences, and becomes sluggish, because the blood flows slowly to it. The whole system is, as it were, partially clogged. The watery fluids, settling in and under the skin, give fulness to the tissues, producing a languid expression, and filling the muscles and lymphatic vessels so as to render them less sensitive to the mental impressions. The movements of the muscles are necessarily moderate because of their bulk, but their size is not an index of their strength.

Where there is an excess of this temperament, the hair is light or pale in color. In childhood it is of a dull white color, but lifeless in its expression. The eyes are tranquil and expressionless. The skin has a dull leady tinge of white, and there is an expression of lassitude in all the tissues. The countenance is listless when at rest. The features are rounded, but elongated, pendent, and heavy, and the lips thick. The secretions of the salivary glands and the olfactory organs, are profuse, and the pulse is slow and feeble. The figure is rather shapeless, and the flesh soft. The body is full and rather thick in proportion to the height, yet there is a general appearance of weakness and apathy.

LYMPHATIC.

BILIOUS.

This is the ordinary phase of the Lymphatic temperament in men. There is a modification of it in the constitution of women, in whom the lacteals are full and active compared with other parts of the Lymphatic system, and this feminine phase of the temperament gives the same placidity and temperamental impressibility, without the heavy, dull, expressionless character above described. In a minor degree, the same modification is observed in children and sometimes in men.

Persons of excessive Lymphatic temperament are temperamentally disinclined to mental or muscular exertion. Nature's way to wake up the system from this inactivity, is by stimulating the action of the lungs or brain. An increased circulation, or such sensations as quicken mental action, the pain of hunger by fasting, for instance, tend to regulate and correct this inertia.

THE BILIOUS TEMPERAMENT.

The Bilious temperament has its centre or source in the liver, which is the secretory organ of the prepared liquids for the nutritive supply of all the tissues, thus immediately affecting both the organization and the functional operations of each of the other temperaments. This organ has been called by physiologists the chemical laboratory of the human economy. Although it is said to evolve more heat than any other organ, its function in this respect is not to increase but to equalize and restrain the animal heat of the system, by its peculiar densifying and secreting process; and this equalizing influence serves as a counterpoise to the heat resulting from the circulation of the expanded oxygenated blood of the arterial system. While the Sanguine temperament tends to a high temperature, the Bilious tends to produce a colder nature.

In persons of this temperament, the hair is black or dark, strong, and abundant. The complexion is sallow, and the skin dry and of an olive color. This temperament gives, also, blackness to the pupils of the eyes, and the general expression harmonizes with their hue.

To take cognizance of this temperament, we must, therefore, observe chiefly the extent to which its indications are apparent in the skin, the hair, and the eyes, and in causing a staid, cool movement in the expansive and contractile operation of the lungs and stomach. It is this temperament that furnishes the solidifying and densifying tendency; and it is the Nervous temperament, acting on this temperament, which gives fineness and delicacy of expression and hue in the skin. The liver, it should be observed, acts by secretion, with functions having relation to latent or low temperature, and more disturbed by high temperature than those of any other organ, and it is more remote from the brain, and more secluded than the organs of circulation from all atmospheric influences and from contact with any thing external. Its action is continuous and steady like that of the brain, and not spasmodic, or expansive and contractive, like that of the lungs and stomach. The liver nourishes the forces of the body, and its quiescent operations are continued without cessation, and are increased and promoted by passivity of the body. The brain, which expends the forces of the body, requires rest, and in sleep its operations are more or less suspended.

SUBORDINATE RELATIONS OF THE TEMPERAMENTS.

In all these general remarks upon the temperaments, I have attributed their peculiarities to the influence of the individual organs, from which they respectively arise. To complete our knowledge of the subject, we ought to understand that each of the leading organs referred to is connected with an associated apparatus, upon the development and condition of which, much of its efficiency depends. The cerebrum in this way stands connected with the cerebellum and ganglia at the base of the brain in the medulla oblongata, and the system of nerves and ganglia which extend throughout the body. The lungs are in like manner connected with, and dependent on, the arterial and venous channels and the action of the heart; the lymphatics and the stomach, on the other parts of the digestive system; the liver, upon the organs subordinately connected with its functions.

The condition of the associated apparatus exerts an important influence upon the leading organ, and may modify the exhibition of its temperamental character. In general, however, the size and activity of the auxiliary apparatus correspond with those of the leading organ, and the phases of the leading organs may be taken as our guide in ordinary observations for the purposes of educational science.

The muscular and osseous systems, (which are often spoken of as the athletic temperament,) and the reproductive system, are of minor importance in reference to mental disposition.

Amativeness is not a *mental* faculty, but the mind often excites it, by the action of the senses of sight, hearing, and touch, as is often the case with man, or the activity of the faculties of Philoprogenitiveness, as is commonly the case with woman; and the mind is often exercised with it, to secure its gratification. In itself considered, it is a physical sensibility, not a mental affection. The cerebellum, the organ of this impulse, is also the organ of unconscious muscular motions.

It has been supposed that the organ of this faculty would afford the best means of testing the Phrenological doctrine, because of the facility with which the observer individualizes both the faculty, and the cerebellum, which is its organ. But much controversy has arisen on this point from not understanding its character as above described. Properly understood it is in itself a physical sensibility, all engrossing in its activity, and when aroused it preoccupies the nervous forces, and proportionally supersedes mental functions. Man was subjected to these conditions by the fall, and Christianity should restore the supremacy of Spiritual conditions.

The so-called "Vital temperament." Temperamental Peculiarities in Childhood.

Phrenologists speak of a "Vital temperament," a "Mental temperament," &c.; and they might, with equal propriety, designate as a temperament, every condition which affords a distinctive contrast between extreme physical or bodily manifestation and the brain.

What is called the Vital temperament, is the phase of organization distinguished by a fullness in the base of the brain and in the neck, where the nerves of sensation and motion from the spinal cord, the great sympathetic, the pneumogastric nerves, the nerves from the organs of special sense, pass in through the various orifices in the walls of the cranium to their termination in the base of the encephalon; that is to say, in the lower part of the Intellectual group and of the Propensities, and in the cerebellum.

When the Lymphatic and lacteal system of the body, in connection with the senses which communicate with the mind, is active, it gives a fullness in the upper part of the neck at the base of the brain, as contradistinguished from the cerebellum, which is not developed until the age of puberty. To this organic appearance no anatomical definition has been given by Phrenologists. The brain itself might not be large in its structure at the base, and still show a great amount of circulatory supply and support of Physical life in this sensuous part, without being at all mentally disposed. In fact, it is demonstrable, that the life of the infant centralizes in this sensuous combination of the different nerves that combine and communicate with the encephalon, *viz:* those of sight, hearing, taste, touch, and smell.

The organization thus described cannot properly be denominated a temperament, but simply the phase of sensuousness given to the mental or physical nature by predominance of vital forces residing at the base of the brain in the nerves of sense, and the combined temperamental influence of the stomach, lungs and liver, and their respective auxiliary apparatus, when it is centralized in this connection.

In a transient form this phase is characteristic of children in health, appearing, however, rather in functional activity resident in the base of the brain, than in well developed fullness.

JUVENILE PHASES OF THE TEMPERAMENTS.

In children, on account of their growing constitution, the channels of sense, sight, hearing, touch, taste, smell, by which the body and mind are both affected, are very active, especially affecting the body, through the nerves of sensation; and these senses, and the desire and necessity for food, stimulate all the conditions of bodily development.

Hence, as a general rule, attention to this sensuous disposition of the child is the direct and immediate means through which impressions are to be made by the teacher. Teachers are prone to disregard this, because in adult life these sensuous external influences affect the mind more largely than the body, and as we grow older we become more obtuse.

Manner in which the temperaments may be described in the combined form.

A knowledge of the four temperaments which I have above described and of the individuality of each, and their reciprocal relations, and of the tendency of disease to centralize itself in one or the other of them, and of the commanding influence over them, ordained to be exerted through the brain, is of greater importance to the curative arts than physicians have generally regarded it.

2. "*If Combinations are to be Regarded, is the Prevailing Temperament to be the Guide?*"

I answer that it is so to be. As I above remarked, every person possesses, in some degree, all of these four primary temperaments,—the Nervous, the Sanguine, the Lymphatic, and the Bilious,—combined; and it is only when one is so predominant that the others are relatively uninfluential in characterising the person, that we may say the person is of one of these primary temperaments. In general, they are so combined as that two or more are well marked, and in some cases all the four appear blended in their due relative order. In the classification of pupils proposed by me in my former letter, in which I advise that the pupils of a class be arranged in four divisions according to these four temperaments, I mean that those in whom a given temperament predominates are to be placed together.

Whenever it is thought practicable to carry this classification a step further, the children of the Nervous temperament should be subdivided according to that temperament which predominates next after the Nervous; those of the Sanguine should be subdivided according to the temperament which predominates next after the Sanguine; and so on.

In describing these combined temperaments, I shall treat the Nervous as the most important element, and shall make it the starting point in delineating the combinations. Not only is this temperament characteristically the predominant element in the American people, commonly leading all the others in the combinations presented, but it is the commanding element in the physical nature; and in the brain, the principal organ of the Nervous system, reside those Spiritual Faculties which enable man to receive the influence of the Holy Spirit, which constitutes the characteristic, distinguishing mark between man and the brutes.

The indications as to which of the several primary temperaments is relatively the strongest are to be sought in the characteristic shape and habit of the body, in its movements, in the general features of the brain as shown in the shape of the head, and in the appearance of the skin, the eyes, and the hair. Of the three latter, the skin affords the most trustworthy indications of temperament, both in its texture and hue, though the indications thus afforded are more difficult to describe, in writing, than those given by the hair and eyes. Special causes not connected with temperamental qualities frequently modify the color of the hair and eyes, while that of the skin, is rarely disturbed thus, except it be by disease or general derangement, and then it is not likely to mislead a careful observer. As explained in the preceding paragraphs, the Sanguine temperament tends to give a red color to the skin, and blue to the eyes; the Bilious gives black; the Nervous is indicated by white in the skin, and grey in the eyes; and the Lymphatic, by a watery and colorless hue. The Nervous tends to a slender form, and to sharpness in all the features of the head; the Sanguine gives fulness and roundness, especially in the chest; the Lymphatic tends to give bulk throughout the system and a shapeless and expressionless appearance. The Bilious gives concretion and imporosity to the liquids throughout the system, tending to impart density to the whole, and affords those qualities which give the basis for metallic force and endurance. The Sanguine gives heat and impulsive force; the Bilious tends to low temperature and quiescent force. In proportion as the Nervous element predominates, in connection with either of the others, it tends to make the hair fine, soft, persistent in its forms, and sparse.

In speaking of combined temperaments, they are designated by combining the names of the primary temperaments, in the successive order indicated by the comparative strength in which the primary temperaments appear united, thus pointing out their respective relations.

When the combination is characterized by a great predominance of the Nervous and the Bilious compared with the Sanguine and Lymphatic, it is designated as the Nervous-bilious. In this case, if the Nervous predominates over the Bilious, there is a bright, brilliant skin, with, however, a slightly sallow tinge, except around the forehead, where the activity of the brain causes a whiteness; the eyes are grey, and the hair black and very fine and sparse, and through all this, shines a mental vividness. If the Nervous and Bilious are about equal, the clear brilliant skin is without sallowness, the eyes are black or grey, and the hair black and fine. If the Bilious predominates over the Nervous, in which case the temperament is designated as Bilious-nervous, its characteristic color shows itself more in the tissues, evidenced by the skin being sallow, and having a passive expression, and by the pupils of the eyes being more dilated and having a colder expression, than when the Nervous predominates.

When the Nervous and Sanguine elements are very much stronger than Lymphatic and Bilious, the temperament is called the Nervous-sanguine; and it presents the colors of red and white, instead of the black and white of the Nervous-bilious. The Sanguine distributes the color of red through the white of the Nervous. This combination presents a well-formed muscular system; the Nervous gives length of fibre and compactness to the muscles, and the Sanguine, by a vigorous pulsation and circulation, gives fulness and roundness, and a pleasing contrast of red and white in the skin, and blue in the eyes. It is to these elements, that grace and beauty of form are to be attributed. If the Nervous is greater in strength than the Sanguine, it gives a clear skin, and the redness of lips and cheeks is well defined, contrasting with the surrounding white, and giving beauty in color.

The Nervous-Lymphatic Temperament. The Bilious-Lymphatic.

The hair is fine and of a pale-looking red color. The eyes are blue. If the Sanguine is about equal in strength to the Nervous, there is a more diffusive expression of the color of red in the whole skin, the hair is red, and there is a lively and warm expression in the countenance. The eyes are clear blue. If the Sanguine is stronger than the Nervous, making the Sanguine-nervous temperament, the same characteristics are increased and intensified, and a high degree of heat is apparent throughout the whole expression of the body, which is like fire in its movements, volatile and quick.

In the Nervous-lymphatic temperament, the skin is rather of a dingy color, watery and lifeless in its expression; the eyes are of a watery grey or leady hue, and are animated only when the mind is exercised; the color of the hair tends to the same dull hue; the countenance, especially the forehead, has a marked whiteness. If the Lymphatic is about equal with the Nervous, more tone appears in the color of the skin, hair, and eyes, than when the Nervous preponderates over the Lymphatic. If the Lymphatic exceeds the Nervous, there will be a fulness in the region of the stomach and in the glandular portions of the countenance, and the muscles of the chin and mouth will sag downward. There will be a watery appearance in the eye-lids, and the eyes will be expressive only at times, when animated. In children, the Lymphatic system is commonly the most active; and hence result free discharges from the eyes and nose, and from the salivary glands of the mouth. The same thing in adults indicates the predominance of the Lymphatic element. In the Nervous-lymphatic, the color is less negative, in proportion as the lymphatic predominates; but there is no very positive expression of color. In proportion as the Sanguine is added to the Nervous and Lymphatic, the color becomes more positive, showing a more decided tinge of red in the hair and complexion, and of blue in the eyes. If then the Bilious be added, it gives deeper tone to the colors; and if it be increased so as to exceed the Sanguine, it gives softness and a deep brown color to the hair.

In the Bilious-lymphatic, there is a white, lifeless, watery expression in the skin, with black eyes of a dilated and blank expression, and the hair is black, and naturally tends to coarseness. In the Bilious-sanguine temperament, the skin is of a deep brunette color. The hair is a dark brown.

When the Nervous has about an equal share, in connection with the Bilious and Sanguine, it gives an expression of brilliancy to the brown color of the complexion, the hair, and the eyes, making it lighter than it is in the Bilious-sanguine.

If with these three elements, the Lymphatic is present in due proportion, then size and fulness of body are given, making the muscles, glands, and the whole contour of the form, well filled and rounded. If, now, among the four temperaments thus combined, the Nervous somewhat predominates, we have all that constitutes strength of muscle, density of bone, and continuance of powers, as well as beauty of the whole body, both in size, symmetrical structure, brilliancy and color of skin, and grace of movement and expression.

What is the most favorable Temperament. Mental Capacity depends on the Brain.

It is easy to see how great is the variety of these combinations, but the foregoing remarks will suffice to show the external indications of relative strength in the combined temperaments. The true order of their relative strength, for the best conditions of life in the temperate zones and in civilized society, is, first, the Nervous, for mental conditions, it being expenditive of force; second, the Bilious, for endurance and tonic support in the system, this being recuperative and unspasmodic in its operation; third, the Sanguine for warmth and geniality of action; and fourth, the Lymphatic for equalizing, and for the necessary supply of liquid for harmonizing what may be called the vegetative conditions.

In general, it may be said that this order affords the highest conditions for health, longevity, and the progress of civilization.

For special vocations, however, some modifications of this order are more appropriate. The Nervous-bilious temperament is for educational purposes and influences the most favorable. This is because these temperaments are not pulsating or alternating, but are quiescent in their functional operations. But if the functions of the nervous and bilious systems exert too controlling an influence, and are too much stimulated by the processes of the school, growth is checked and decrepitude results, for want of the healthful influence of the vivacious alternating action of the lungs and stomach. Hence, active out of door sports should be particularly encouraged in children of this class.

DISTINCTION BETWEEN MENTAL AND TEMPERAMENTAL QUALITIES.

One great difficulty in carrying this method further than the four-fold classification, heretofore suggested, is found in the fact that while many physiologists are agreed in recognizing the four leading temperaments, substantially as above described, they differ widely in regard to the exact relation which is marked by each, with and upon the mental conditions of the brain, going so far as to characterize them as the poetical temperament, the musical temperament, the sensitive temperament, &c., and thus confuse the whole subject of physiological investigation.

A clear knowledge of character cannot be attained without a careful discrimination between those elements of mental capacity which depend upon the organization and development of the brain, and the organs of the special senses, and the characteristic grouping of the faculties, and those elements of force or quality of action which depend upon the bodily temperaments in connection with the sensuous activity.

To illustrate this distinction by an actual character, I give, in the diagrams, the head of George Washington, whose character is more familiar and justly more admired and venerated than that of any other person named in our annals; and it will be seen how clearly the great attributes of his mind are defined in the combination of temperaments, and the marked form of the head. Every intelligent observer, looking upon it, is impressed with the meditative placidity, the dignity, and the might of the character portrayed in it; but not many can give a definite statement, beyond a reference to the law of suggestion, why such lines and shadows should convey the image of such a character. Physiology and the science of the mind enable us to analyze the material form, to trace each physical peculiarity of color, texture, and shape, to its co-related physiological and phrenological causes, thus unravelling the connection, and estimating the relative force, of the great functions of the system, and of the groups, clusters, and special faculties of the mind, justifying, but far transcending, in the exact and detailed result, the first, vague and undefined impression.

In deducing a practical method of analyzing and demonstrating character, from the principles that mental nervous function is localized in the cerebrum, and that local development in size results from activity, Phrenologists have assigned fixed localities for the respective convolutions, and have given definitive descriptions of the functions they ascribe to these organs, without sufficiently regarding the differences observable in both these respects, in each and every head. It is only in a relative sense that a faculty can be said to occupy the same place in different heads. The convolutions corresponding to the respective faculties appear to have, in all heads, the same general position among themselves, when relatively considered; but such is the endless variety of form caused by different development, that the location of many convolutions, when compared with the general outlines of the head, is very variable. Correspondence to any ideal location, is rather the exception than the rule. Thus, we cannot assign to Destructiveness, for instance, a uniform position. In some heads, this faculty appears in its appropriate central position, in the space immediately between and above the ears; and the centre of its location in such a case, would be indicated by the vertical line h A, (fig. opp. p. 63,) rising from the centre of the opening of the ear. In others, it is developed backward, so as to appear behind that line; in others again, it is developed more forward of that line; and in others still, a downward development gives it a position between the ears and extending both in front and back of them, bulging them out. These diversities of location, and the variations of associated action, and of physiognomic expression, connected therewith, are significant of corresponding diversities of mental character. If one head is found to be higher, broader, rounder or fuller, or phrenologically more angular, than another, it is because some parts of the brain have a marked development in a certain direction, necessarily giving some of the convolutions a location removed from that which they otherwise would have had; thus altering distances generally, though not necessarily disarranging relative positions. Hence no two heads resemble each other in the precise location of the faculties, if compared with an absolute standard, any more closely than they do in the general exterior form.

The diversity in this respect may be compared to the diversity which appears in the location of the features of the face. The relative contiguity of the features is always the same; but if two faces are measured by lines such as those shown in the diagram, the mouth, which in one will be seen

Diversity in the Associate position of the Organs. The mind not to be arbitrarily measured.

below a given line, will appear in another to be above the corresponding line. In one, the mouth might appear to be wholly within one square, and in another, it would extend into three squares, and perhaps rise into a fourth. In one, the eyes being small and close together, would be seen in contiguous squares; in another, being large and far apart, they would chiefly appear in the squares further removed on each side. Thus, although in the general relative position of the organs, all heads are alike, yet when the location of the organs is accurately defined, each head has its own proper individuality of phrenologic form, in which it is unlike any other.

No chart or bust, therefore, can give anything more than an ideal or average shape of the head, and an ideal or average position to the faculties, unless it be the representation of an individual head like that presented in the diagram. This gives the location of faculties in the case of this particular person; and enables us, by contrast and comparison, to describe certain characteristic variations of position indicated by other forms of head.

Again, not only do the locations of the convolutions or organs vary, but the posture of each towards others, and the degree to which it is distinguished in form from them, or merged in them, are equally variable. For instance, the organs of Destructiveness and Secretiveness usually are distinguished; but the convolutions are often so associated in position and activity that in exterior form they show but one development. This combination manifests a wary executive force or strategic power. If the development of Destructiveness, is on the other hand, not toward Secretiveness, but downward on the side of Alimentiveness, there will be a manifestation of executive power in the sensuous passions. If, again, the Intellectual group predominates, and Destructiveness is developed forward and upward upon the side of Acquisitiveness, there will be executive power intellectually.

We cannot estimate the mind by any arbitrary scale of mensuration of its organs. We are surveying the most complex and wonderful of the works of the Creator; the only creature in whom spiritual life and physical forces are combined; the organized being in whom alone the Almighty Creator has embodied His own image, and whom alone He has gifted with capacities to receive His own direct guidance. It is to little purpose, that we frame artificial formulas of comparative bulk and manifestation, and balance the results to find the sum of the mind. The Nervous forces refuse to be measured by a process which takes no cognizance of the most essential elements of the problem. We must ever remember the infinite diversity of form and appearance which growth manifests; and in seeking for the significance of these external manifestations, we are to proceed by a knowledge of the laws by which these functions are determined. This must include both a recognition of the plastic power of organic nature, which eludes artificial and absolute measurement, and also those Spiritual laws I have referred to, which, when brought into operation in the soul, modify, regulate, and illuminate the organization. The discernment of this Spiritual life is the first condition for knowing the character of another who is the subject of its influence; but "The natural man receiveth not the things of the Spirit of God, for they are foolishness unto him: neither can he know them because they are spiritually discerned."

In presenting the outward definition of the mind, by its organic form, depicted for the Perceptive faculties, as I do in these pages, I point to the visible objective manifestations of the natural faculties. The life of the Spirit is only known by the Spirit.

The faculties of Godliness, with their six pair of associated members in the Spiritual group of faculties, constitute, so to speak, the crowning centre of the mind; and by the Divinely ordained law of the nature of man, it is only by the sentient power manifested through this superior part of the organization that the soul has its true life; and to know this spirit in another the observer should possess it.

It is only by bringing to the analysis all that is known of the laws of growth, and looking to the general form of the head for the cast of mind, and thence proceeding to ascertain the special characteristics, that we may analyze the objective facts so as to form a just perceptive judgment of the character.

Before the phrenologic observer directs attention to the form of the head, he should, therefore, observe the temperament of the person; for the peculiar combination of the temperamental elements presents the background upon which the mental character is to be portrayed by Phrenologic analysis. Having thus observed whether the brain predominates over, or is subordinated by the other three temperaments, we must observe in what order and degree the body and mind are sensuously connected and blended. In doing this, the superficial observer is prone to be misled by mere regard to the physiognomic expression, particularly by that of the eye. But the development of the base of the brain and the neck, and the physiological acuteness and quickness of the senses are the true indications of this characteristic. Having thus learned the tendency of the mental life, whether outward and demonstrative, or interior and self-involved, we are prepared to observe next, the general form of the head. By ascertaining the order of development in which the three groups of faculties stand, in that form, we shall ascertain what is the general mental disposition, that is to say, whether the mental life, manifested under the influence of the temperament, and in the immediate sensuous action, is characterized by a predominance of the Animal and Social, the Intellectual, or the Spiritual group. By the knowledge of these general characteristics, the observer is prepared to distinguish the clusters and families of faculties from each other, and to recognize the location and development of particular faculties; hence ascertaining the peculiar qualitative character imparted to the whole disposition by the deficiency or predominance of any one.

The influence of the temperaments upon mental disposition, thus observed, may be illustrated as follows.

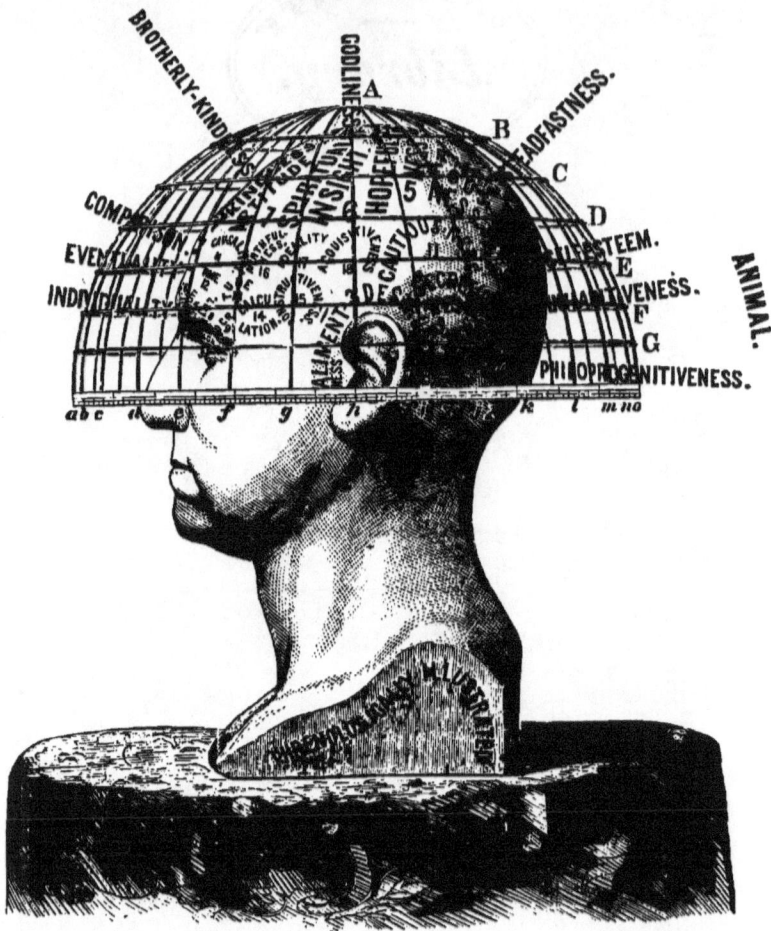

PHOTOGRAPHIC VIEW (CUT IN WOOD) OF THE PHRENOLOGIC BUST OF WASHINGTON.—No. 6.—*Left side*.—Full profile view representing, as far as practicable, the special Faculties of the three groups. Upon this side of the bust, the locality of each special faculty is designated as it was manifested in the head of George Washington. The different size of the lettering employed for the faculties of the different groups indicates the characteristic difference in the size of the convolutions. The regions of the groups are indicated, in the cut, by their names in the margin.— See pp. 16, 30 60, 109, 111, 112, 115, 118, 147, 148.

[To face p. 63.]

OF EDUCATION. 63

Indications of Mental Character in the side view. Diversities of temperament.

In the profile view of George Washington, the part of the brain which appears predominant is in the group of the Spiritual faculties, comprising both the Meditative and the Intuitive cluster. It is evident from the shape and bulk, so far as they are shown in this view, that the Meditative part largely predominates over the Intuitive part; and that the Intuitive part is drawn and subordinated to the Meditative part. So also is Self-Esteem, which thus consorts with Steadfastness. This form indicates that Premeditative feeling, rather than Intellectual thought, was the source of his actions.

This fact, in connection with the qualitative influence of the temperaments, would give the observer one chief element in the analysis of the mental character. Thus, if the Bilious temperament predominated over all the others, the tonic strength of its influence, with this development of brain, would give austerity and stiffness, tempered with sufficient aspiration and reverence to impress an observer with the idea of commanding qualities. If this had been the temperament of Washington, jealousies and suspicions would have resulted in the minds of those about him in reference to his plans, from his deportment, and in consequence of his youthfulness; and from the crude state of the political elements, such jealousies would have probably precluded success.

If the Nervous exceeded the Bilious, both predominating over the other temperaments, there would have been a more active mental force; and, since development of the Meditative cluster is the condition under which judgment of a higher quality is manifested when its powers are called into exercise, the Nervous-bilious temperament would have made him superior, mentally, on account of the activity of the brain, and superior as a judge, on account of the reflective power of the Truth. If this had been the temperament, it would have given him a greater pre-eminence, by reason of the greater mental force; but he would have tended to increase the distance between himself and the men who gave him their support, who would have been alienated by his being not sufficiently accessible, while the jealousies of rivals and the opposition of those who objected to placing him in command because of his youth and inexperience, would have been increased in consequence of his greater mental activity.

If the Lymphatic temperament preceded the others, they following in the order above indicated, it would give a more passive character, with the reticent influence of the Bilious; and he would not be so ordinarily or easily influenced by surrounding circumstances. Bodily exertions would be distasteful, and the mind would be characterized by inertness. With a marked degree of the Lymphatic temperament, as compared with the others, the body would seem to take precedence of the mind, and the man would be lazy. If now, while the three temperaments we have named preserve this relative order, Bilious-Nervous-Lymphatic, the Sanguine were given predominance over them all, the man would be more vivacious and fitful; and although he would still be unwilling to act without necessity, yet when under necessity, he would show a great deal of passional force. With this disposition he would be an active leader, working in the immediate sphere of Social and practical life.

Turning now to the front view, (No. 8), which is the next in importance. we shall see in *this* general form of the head another element of the analysis of the mental character. In this view the height of the head is not so apparent as in the profile view, because of its great breadth. This breadth, in the position indicated in the diagram. and in connection with the influence of the temperament, shows the amount and quality of vital force residing in the executive and restraining faculties. The breadth, in this head, is greatest at a point above and behind the ears. If this extreme breadth were at the base of the brain, it would show a downward development and a gross, sensuous nature. If it were in the back of the head, it would show great social force in the Propensities located there. If it were in the forehead, it would show great Combinative capacity of intelligence in some form. Being, in this head, in the faculties of Destructiveness and Cautiousness, and upwardly developed towards the other restraining faculties, Righteousness and Steadfastness, it shows the great amount of vital mental force latent in the executive and restraining faculties of the Propensities. The commanding mental qualities of Steadfastness with both Godliness and Self-Esteem, indicated by the form, in height. in the profile view, in connection with those of Executiveness and Cautiousness indicated in this form of breadth seen in this front view, so far as mental conditions go, fitted Washington, so far as the organization and order manifested in Phrenologic form is concerned, for high faith, executive ability, and more than equal latent power, in the great cause in which he was engaged, and constituted him the great moral leader of his people.

This additional view of the form of the head enables us to enlarge our statement of the temperamental influence as follows :—If the Bilious were predominant, it would give self command ; the man would feel superior to the conditions by which he was surrounded, and, although not possessing philosophic conception, would manifest great practical forecast. In the case of the Bilious temperament, there would be, with the bland austerity, indifference even to immediate surrounding circumstances; so that the mind would seek retirement, and it would require some great necessity to call forth the full powers. In ordinary circumstances, the activity of the mind would be self-contained and undemonstrative.

In the case of the Nervous-bilious temperament, this breadth through the restraining faculties would indicate, with the judicial and social power, a watchful and anticipatory attention and adjustment to the immediate conditions of success, in all their details, bringing the mind into more intimate relations with passing events ; and, with the strength of perceptive faculties which Washington possessed, which may be seen in the sharp outline of Individuality and Eventuality, success would attend effort,

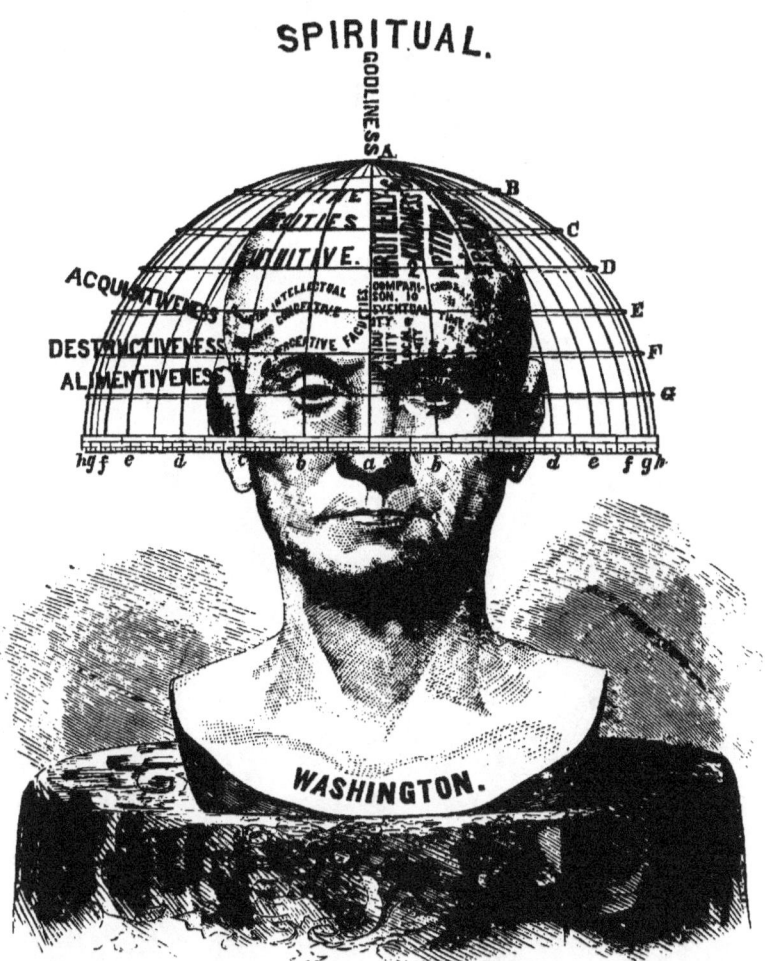

PHOTOGRAPHIC VIEW (CUT IN WOOD) OF THE PHRENOLOGIC BUST OF WASHINGTON.—*No. 7.*—*Full front view.*—Representing on the left side the names of the Faculties, and on the right side the location of the groups.

In this view is shown the locality of the Perceptive, Conceptive and Combinative clusters, respectively, which together form the Intellectual group. In the upper part of the head the region of the Intuitive Spiritual Faculties in both hemispheres is prominent in this view. –See pp. 111, 112, 115, 117, 118, 119, 136.

[To face p. 64.]

PHOTOGRAPHIC VIEW (CUT IN WOOD) OF THE PHRENOLOGIC BUST OF WASHINGTON.—*No. 8.—Full back view.*—Showing on the left side the names of the Faculties, and on the right side the location of the groups.

In the lower part of the head the region of the Propensities in either Hemisphere is shown. In the upper part is the region of the Meditative Spiritual Faculties. The lettering which indicates the general region of the groups is prominent on the right Hemisphere, while the names of the special Faculties are indicated on the left Hemisphere, Acquisitiveness, Destructiveness and Alimentiveness projecting in a direct line from the surface of the head, and their location designated.—See pp. 115, 128, 131.

[To face p. 65.]

| The character of George Washington described, Temperamentally and Mentally. |

because of the constancy of watchful attention to apparent facts, the superior judgment, the practical activity, and the strong force of the character. This does not describe the character manifested by Washington under ordinary circumstances, because the predominance of the Lymphatic temperament made him indifferent to immediate influences until the greater necessity presented the moral obligation. If the Sanguine should predominate, the man would be more affected by immediate circumstances, and would show by his life, warmth of affection and social interest for those who surrounded him: his happiness would depend upon his family, and his mental life would be more made up of the immediate, present, surrounding influences, and of action and re-action between the mind and external affairs. He would be eminent in athletic sports, on account of the large preponderating force indicated by the breadth between and above the ears, in connection with the elements of character shown by the other view, and the combination of the temperaments. If the Lymphatic should precede all others, which was the case with George Washington, especially in the later years of his life, (while in his earlier life the Sanguine had more influence, and in the middle the Nervous had more, when great necessity was upon him,) the tendency to inactivity indicated by this breadth of head would be increased, and this disposition would be overcome only by the strongest appeal, such as his country's call was to George Washington.

In Washington's case, it is probable that if the events transpiring had not been of so momentous a character, he might have died in obscurity. Out of his temperamental weakness, and the peculiar order of his mental faculties, the irresistible necessities of his country brought strength. The fitness of Washington for the task imposed upon him may be concisely stated thus:

He possessed great executive force, (shown by the breadth of head,) and remarkably clear and single perceptions, (shown by the outline of the forehead in the profile view), but these powers were rendered inaccessible to ordinary selfish influences by his passive temperamental conditions. Only through the great sense of moral obligation given by his meditative faculties, (the predominance of which is shown by the height of the head in the back part,) could this latent power be called forth ; and hence the necessities of his country alone could command him.

The change in the character of Washington was marked by his temperamental development, as he advanced in life. In early manhood his hair was sandy inclining to yellow; this appearance indicating the warmth of his interest in affairs at that period of his life. The infusion of this vivacious, Sanguine temperament gave him his disposition to delight in athletic sports. This warmth of disposition in his earlier years, made him genial and pleasant in his social relations with others. Being a temperamental rather than a mental disposition, it naturally had its most sensuous, and freest manifestation in his private life ; while in the duties of office and military command, his mental disposition led, and his high qualities of Righteousness, and Steadfastness gave him a colder and graver tone. The same warmth of temperamental disposition was manifested also, by the high spirit which he exhibited when unjustly treated in the military appointments made by the Colonial Governors, a spirit which contrasted very strongly with the disposition of his maturer years, when placed under similar circumstances. As Washington passed into the mature period of middle age, the influence of the Sanguine temperament declined, and that of the Nervous increased, and as he entered upon the period of the Revolutionary war, the Lymphatic temperament declined, and the Nervous increased still more, until it preceded all the others. The order which characterized the combination of the temperaments which he possessed at this the most active period of his life, was the Nervous-Lymphatic-Sanguine-Bilious, the Nervous and Lymphatic elements being much more influential than the others, and the Bilious having no appreciable effect on his mental disposition. His hair had a tendency to become white and quite sparse before he attained old age, indicating the necessitated activity of the brain. In the later part of his Presidential service, and in his subsequent retirement, the Nervous again declined, and the Lymphatic increased, returning to its early predominance; but the low degree of the Sanguine and Bilious elements continued to his death. The natural predominance of the Lymphatic temperament in him, the declension of the Sanguine, and marked deficiency of the Bilious, give the Physiologic expression of his characteristic disposition;—thoughtful, imperturbable, unimpulsive, and unambitious.

Each of these temperaments have their epoch in human life, growth usually commencing with predominance in the Lymphatic, next developing the Sanguine, the Nervous coming to its highest power in maturity, and the Bilious, which is always the underlying base, giving contrast to the other three, both substantially and phenomenally, often being more prominent as the others decline, towards the close of life. Variations in this order exist in different individuals and nations. In Washington, as in many others, the Lymphatic continued to increase with age. The shifting combination marks its influence in the changes of mental character.

IMPORTANCE OF TEMPERAMENTAL OBSERVATION.

From these illustrations of the influence of the temperaments, it will be seen how important it is that the observer should first possess himself of the temperamental peculiarity of the subject of observation, as the basis of Phrenologic estimates.

A knowledge of the temperaments will tend to lead the mind of the teacher to a sense of the necessity of Spiritual guidance. His is the high function of directing in the formation of character; and yet, when he is dependent on his own natural mental force, and subject to the full natural degree of the qualifying influence of his own temperament, he finds him-

Deficiencies in Phrenologic systems in respect to Temperament.

self constantly experiencing a painful want of adjustment between his own powers and the subjects of his effort. It is only by that humble and earnest spirit which the predominant activity of the Spiritual faculties, when the soul is under the influence of the Holy Spirit, gives, and the resulting power to modulate the manifestations of his own special temperament, that he can rise to the vantage ground which he must attain for the successful accomplishment of his task. When the activities of the mind, and the influence of his temperament are thus harmonized in the Will of God, and not otherwise, can he exercise the full power which the law of his being affords him.

The observer having estimated the temperamental conditions affecting the mind, he is prepared to measure the relative development of the groups. A due regard to this essential three-fold division of the mind is of fundamental importance, both in Phrenologic observations, and in dealing with mental processes. It is because the qualitative characteristics of the Spiritual part, as compared with the others, have not been recognized by Phrenologists, that they have not succeeded in establishing the Phrenologic doctrine beyond controversy. They have investigated the facts of physical organization, and the corresponding manifestations of faculties, and found confirmation for their conclusions, in Comparative Physiology. Dr. Gall presented the great truth of the mental significance of peculiar conformation of the head, and elucidated the character of many of the phrenologic features or regional developments, nick-named bumps. Hence the science as presented by Dr. Gall, is more appropriately termed Cranioscopy. Dr. Spurzheim, without recognizing the special mental quality of the three groups, extended the observations which Dr. Gall had commenced, by carrying the individualization of the faculties to a greater detail, thus presenting a minuter division. His labors, by attempting a more systematic analysis of the mind, presented the subject in the aspect of Phrenology.

Phrenologists in seeking to define the exact location of faculties by superficial measurement, and endeavoring to reduce the results of observation to mathematical formulas stating the quantative predominance of the faculties according to a numerical scale, have presented an arbitrary and artificial view, inconsistent with the plastic and interblending forms manifested by all organic growth.

Phrenologists have heretofore not rightly apprehended the general division of the mental qualities of the brain, and therefore could not give a proper analysis of the mind and its qualities. They have known the Intellectual Faculties, and the Social and Animal Propensities and Moral Sentiments, but have not known how to define the Spiritual, the Meditative and Intuitive feelings. They have recognized only that passive existence which the Spiritual Faculties have in connection with the Social Propensities; but have not recognized the activities of the Spiritual Faculties, which the direct impartation of the Holy Spirit of God induces.

The Temperaments present the Organization, only; the Soul is undefined.

Physiologists have generally failed to understand the relations of each temperament to the others. They have studied what have been termed the vegetative functions, to the disregard of the mental. They either have wholly omitted mental conditions, or have given them no proper form and order. Many have given more credit to the physical functions, for influence upon mental conditions, than they ought to receive. And some have regarded the temperaments as the conditions of the mental disposition; whereas the primary conditions are in the brain, (the Spiritual faculties being designed to be predominant among these conditions,) and the temperaments, in their combination, act reciprocally under the brain, being the recipients of the mental forces.

These four colored portraits present the phenomenal aspects of the describable outgrowths denominated temperaments, centralized in these four characteristic qualities, and delineated so far as the contrasts of color present them in the appearance of the head, so that the perceptive faculties, by Individuality may clearly discriminate them. As far as physiologic knowledge admits, we have described their objective manifestations, indicating the influence of each in the phases of human life, and have pointed out the functional seat and centre of each.

Thus far we have been treating of that which is visible and tangible, and therefore capable of description.

We must always distinguish between these objective facts, of which the senses take cognizance, and that indescribable something called the soul, which can be observed only by its manifestations through the body; but of which the fair deductions of proper reasoning, from the premises here stated, give us cognizance. The spirit itself we cannot define. Besides the physical organization which man possesses, he is, by his original creation, or by being born again, a living soul, existing in this respect in the image of God, who is a Spirit. And when we speak of the body or brain being the residence of the soul, we speak only of the essential location, in space, of its centralized, visible, and physical manifestations.

3 "*How are these Distinctions of Temperament to be made Available?* 1. *In discipline;* 2. *In instruction?*

Persons resembling each other in respect to a predominance of the sensuous influence of either of the three more physical or bodily temperaments, tend, strongly and unconsciously, to interblend, sympathize with, and influence each other. Differences of mental constitution modify this tendency and prevent it from resulting in a mental or social affinity, but the similarity in phrenologic form of development promotes the power of sympathetic influence. This is true of classes as of individuals. Where there is a diversity of temperaments, therefore, between teacher and scholar, there is more need of self-adaptation, in order to secure the necessary, interblending sympathetic influence. Mediately, through the agency of the

Holy Spirit this can be made practical; and it will be most practicable when the Spiritual faculties of the teacher have an orderly development, Godliness being predominant, making him first humble.

The first condition, therefore, required to make these distinctions available, is for the teacher to consider his own temperament, in connection with those of the children under his charge. The teacher is always more likely to gain an ascendency over children who are of the same temperament as himself, than over others, especially when the general forms of the heads are alike. Those in guiding whom he will find the most difficulty, will generally be of a temperament the antithesis of his own, especially if the relative order of development of the several groups in the organization of the brain be different. If the teacher observe this fact, he will be better able to adapt himself to the children who are unlike him in temperament, and to modify his mental dealing with them, so that he may interblend his temperamental and mental conditions in the work of education, to conform to their temperamental peculiarities, and thus gain an influence over them, by a subjugation of his own peculiarities.

Where the teacher has in his class those of a temperament different from his own, he will best reach them through the sensuous sympathetic action of those who are like himself. By the interest which he excites in those who are like himself, temperamentally, he may better suffuse the whole class. In attempting this, he may begin by addressing himself generally to the whole class, and thence observe those whom he best reaches, or he may, by observing the temperaments, select his instruments, in the first instance, and address himself to them.

The teacher must, therefore, realize in his own consciousness what is his leading or predominant temperamental disposition, so as to control its exercise.

The second condition requisite to make these distinctions available is, that the teacher should realize how much of the life of a child is, by nature, in the nerves of sense which communicate with all the temperaments, and in the Propensities and the physical or bodily functions, what part is in the Intellectual Faculties, and how little there is in the Spiritual, and why this should be so; and, by a careful observation of the temperamental disposition of the child, he will learn what faculties he must appeal to, in order to get possession of the affections. A child of a Sanguine temperament and active Social Propensities is full of play; and if the teacher would gain ascendency over him he must enter into his plays. This is the way to gain the possession and guidance of those forces which he desires to bring forward into the Intellectual faculties.

In order to do this he must establish the temperamental affinities between them. He must come to the child's condition, and not expect the child to come to his. If he is nervous, or mentally exercised unfavorably toward a child, that child will first imbibe that about which the teacher is thus exercised, instead of what he wishes to teach him; and he will feel in himself a resistance. Perturbation in the master begets the same in the children; and without, knowing it he is doing the very thing he would not, and leaving undone the thing he desires to do. Although the teacher cannot actually modify his temperamental disposition so as to meet that of the child, yet, if he is aware of the diversity of temperament, he can, more especially with a meek and humble state of mind, by self-restraint and control, modulate his expression and bearing to the child, so that the child will be able to receive the impressions he wishes to impart. It is by the teacher's own mental effort, adapting himself to the volatile and lively sensuous disposition of the temperament of the child, and to his playful ways, that this affinity is established.

The first step in teaching is to gain attention.

The Sanguine child will be on the alert, immediately to hear what is going on; and the same temperamental disposition that makes him quick to listen when the teacher speaks, makes him quick to be diverted, unless his mental organization is such that his thoughts are centralized in the contiguous faculties running through the centre of the head, causing his attention to be continuous, or unless his restraining faculties are large, in which case, if the teacher can retain his attention, it will be effective. The Sanguine and Nervous temperaments, being volatile in their nature, are more susceptible of influence than either the Lymphatic or Bilious.

The Lymphatic child is more slow in receiving mental impressions, which are often not distinct nor long retained. If the appeal made is to the faculties of Alimentiveness, his attention is secured at once.

In the Bilious child, if the liver is called into too great activity, the activity of the Lymphatic and Sanguine systems is proportionately checked. The characteristic of this child is inactivity, with a tendency to segregation and quiescence. The impulsive nature of the other temperaments has given way to the passive nature of the Bilious. The instruction which the teacher imparts is not eagerly received, but what is received is more permanent.

The Nervous-bilious child does not give attention as quickly as the Sanguine; but when his attention is gained it is more likely to be continuous, for the habitual exercise of the mental faculties is pleasurable to him, and the nature of this combination of temperaments is quiescent and not marked by pulsative and physiologic diversion. If however, this combination is excessively predominant, the liver will be called into too great activity by the requirement of the brain, and the activity of the stomach and lungs—the Lymphatic and Sanguine systems—will be proportionately diminished, and the growth of the child will be checked. Where this predominance appears, great care should be taken to promote active out of door sports, and secure all the external conditions of active vegetative life.

It will be difficult to get the attention, unless the teacher has the affections of the children. The citadel of children's love is the sensous connections or communication of the senses with the brain. The attention having been gained, the next thing to be regarded is so to administer the methods of discipline as to keep those affections, while also exciting a sufficient degree of intelligent Cautiousness to give circumspection and keep the mind on the alert.

The teacher is not to abandon the use of fear, but must rely on it as one of the most important conditions. He is not merely to use it on rare occasions; but, in a proper degree, and subordinate to love, he is to use it continuously. When the teacher has the attention and the love of the child, and, subordinate to that, a sufficient activity of the restraining faculties, Cautiousness, Secretiveness, Conscientiousness, (or Righteousness), Firmness (or Steadfastness,) the intellectual mind is open to him, ready to receive and appropriate whatever instruction he may impart.

From the foregoing description it will be seen, that the mind of the Sanguine child is most readily awakened by a vivacious gesticulation. With the Nervous, acute, mental clearness,—precision and vividness of thought and language must be employed. The Bilious are more influenced by a grave, sedate, and persistent application; whilst the Lymphatic commonly need to be urged by a sharp, spicy method.

At this point, it will be observed how important it is, always to bear in mind the difference between the child, in whom the active development is chiefly in bodily or physiological growth, and the teacher, whose bodily growth is accomplished, and whose mental powers are in full strength and vigor. How striking is this distinction! On the one hand, the teacher, with educated and matured intellect, his force centralized in the brain and capacitated, according to his own peculiar temperamental conditions, to acquire and permanently retain any species of knowledge; on the other hand, the scholar, not fully developed even in bodily strength, and with a temperamental disposition, which if not in harmony with that of the teacher, shrinks from impressions, and a mind, which, like water, yielding to impressions, quickly loses them.

How necessary, then, that the teacher's spirit should be so modulated that his intercourse with his pupils may secure their sympathetic feeling, interblending the influence of his mind in theirs, and not arousing the resisting forces!

Assuming these relations between the teacher and the children to be established, I proceed to answer the above question in its application to the discipline and instruction.

(a) AS TO DISCIPLINE:

1. *What Temperaments are best Treated by Coercive Means?*
2. *What by Persuasive?*

When these two means of discipline are considered upon Physiologic principles, the proper use of each, and the method of its application, become apparent.

There is no particular distinction between the temperaments, as such, in respect to whether coercive or persuasive means should be used.

Different children susceptible to different motives. The development of the Temperaments.

The selection between these is governed by other considerations. But if either is to be used, the distinctions between the temperaments are of great importance, in connection with a knowledge of the mental disposition, in determining the nature and degree which should be used, of either kind. Thus, to require a child to stand still for a certain time, would be a much greater punishment, if he were a Sanguine child, than if he were a Lymphatic or Bilious child. To impose a punishment requiring a considerable physical activity, which will be very oppressive to the Lymphatic or Bilious temperament, may prove a mere frolic to a child of the Sanguine temperament. The same distinctions apply to persuasive measures. The promise of a cookey may be a strong inducement with the Lymphatic temperament, while to produce the same persuasion on the Nervous temperament, the promise of a story or a picture will be more appropriate, and to the Sanguine temperament, the promise of a game of ball. Every teacher observes these differences in children. What I would point out is that they are chiefly dependent on, and explained by temperamental conditions, and that a classing of children by temperaments, and a due regard to the affinities of the mind directly affected by the temperamental conditions, would very much facilitate the work of instruction and diminish the necessities of discipline.

If the teacher finds a Lymphatic boy dull and stupid, he should ask if the mother gives the boy all the food he wants, and if she does, he should ask her to diminish his allowance. The abundance of food engrosses the activity of the system in the stomach, and the child cannot learn his lesson until he has digested his too hearty meal, or if he does, derangements of the stomach will result. Instead of whipping the boy for his dulness, the mother should be required to diminish his supplies of food before school hours, and then he can learn. If the appetite is restrained and controlled sufficiently to direct the exercise of the physical disposition from the stomach to the brain, the communication being through Alimentiveness, the forces will be directed more to the brain, transforming a physical disposition into a mental one. Thus if the Lymphatic child is made hungry, Alimentiveness in the brain is called into activity, and if the next contiguous faculty, Destructiveness, is large, he will begin to move to supply his want; if the next faculty, Secretiveness, is large, he will be sly to look about him to find how he can help himself; if the next faculty, Combativeness, is large, he will be ready to fight for food. In this way, depriving the stomach of its food, awakens the vital forces which are resident in the propensities. Now, if the deprivation is not so severe as to amount to hunger, the forces of the mind, thus awakened, can be called forth, by the teacher, into the Intellectual Faculties and exercised there. But supplying the stomach with all it will receive, in cases of this temperament, will preclude mental activity. As this temperament is more or less predominant in almost all children, due attention to the food is of fundamental importance, in endeavoring to develop, control, and guide mental activity.

The temperaments are all susceptible of influence by mental and physiological means, and the activity may be stimulated in each, and may be distributed from one temperament to another, and to the brain, as educational processes demand; but children of different combinations of temperament are to be *reached* through different kinds or methods of persuasion, to be chosen with reference to the peculiar mental as well as temperamental disposition of the subject.

The stomach and the lungs, both being alternative in action, harmonize with each other, and either may be readily influenced and stimulated through the other. So the brain and the liver, being quiescent, harmonize, and may be influenced by each other. The brain is more directly dependent upon the liver than upon either of the other two organs, and therefore by a great activity of the brain, the liver becomes torpid.

The direct means of influencing the vegetative temperaments are found in the bodily conditions, respectively, exercise, food, and sleep. Persuasive means, unless the Nervous temperament predominates, are only the indirect means of influencing the temperaments. The difference between different children, in their susceptibility to persuasive means, is partly mental, depending on the conformation of the brain, but in a larger degree it depends upon temperamental character. This difference with adults, is mental rather than temperamental, whilst with children it is temperamental rather than mental, on account of the predominance of the Lymphatic, sensuous, and growing disposition. Under appropriate circumstances, either coercion or persuasion becomes useful as indirect means for securing the attention and control of children of either temperament. My explanations as to the considerations which should determine us in choosing one kind over the other, and as to the best methods of employing them, are reserved until I come to answer the questions in the third division of your letter.

(b) "As to Instruction.

1. *What Temperaments are most Inclined to Study?*

I answer: the Nervous temperament, resulting as it does from the predominating size and activity of the brain, and giving activity and a leading influence to all the mental operations, is the temperament which in connection with a development of the Intellectual Faculties, and with the necessary bodily conditions, characteristically displays intelligence; and this is the temperament which is most favorable to study. There is no necessary difference between the other temperaments, as such, in this respect. It cannot be said that either one, as a temperament, implies any superior inclination to study, with the exception of the remark already made upon the superiority of the Nervous-bilious temperament, nor is it accurate to say that the Nervous temperament inclines its possessor to study. It is not that this

Suggestions as to Modifications in Treatment.

temperament prefers study as such; but that it is the temperament best adapted for any occupations involving or requiring activity of mind. It is upon the Phrenologic development of the brain organs, relatively considered, more than upon the temperament, that the inclination to study chiefly depends. The Nervous temperament gives a general predisposition to mental activity. In childhood, the Animal region of the brain necessarily predominates, by its physiological predetermination, over the Intellectual and Spiritual regions; and while this remains the case, the activity of the mind seeks exercise, not in study, but for bodily development, in other pursuits and exercises, characteristic of childhood. In adult life, if the Intellectual region of the brain has acquired predominance, this activity will predispose the person to take an interest in studies; and if the conditions afforded by the Bilious temperament are favorable, the necessary adherence, and love of study, perseverance, and continuousness of effort are displayed.

2. *What Modifications in Treatment should this Lead to?*

The teacher should first get his mind in immediate contact with the sensitiveness of the children. The Lymphatic children should be seated immediately at hand, near the teacher so that he may more easily arouse them. The Sanguine children may be seated far away, for their attention can be quickly called. The Nervous may be seated next beyond the Lymphatic, and the Bilious between the Nervous and Sanguine. This puts the cooler Bilious temperament next to the volatile Sanguine.

There would be more success in teaching if these four divisions were separated, and each put under a teacher of corresponding temperament. But it should be borne in mind in carrying out such a classification, that the apparent temperamental distinctions are greatly modulated by parental and home influences, differences of mental disposition, constitutional peculiarities, and conditions of necessity, health, food, and fatigue.

The teacher should, according to the foregoing principles, regulate the length of time spent in study and recitation, the methods of communicating information, the means of infusing his subject with interest for the juvenile mind, the degree of repetition and reviewing, the intervals of relaxation, and the extent and variety of recreative sports, as well as the means of discipline and correction, by adapting them intelligently to the particular characteristics of the child and the class of children he has to deal with.

The methods of training should be so modified, also, as to promote an orderly development of the temperamental systems, and avoid or supersede the tendency to extreme predominance of any one over the other. The Sanguine child should not have all the exercise he wants. The Bilious child must not be allowed all the retirement and bodily passivity he may seek. The Lymphatic child should not have all the food and sleep he would take; and the Nervous child should have less studying and thinking than he wishes.

The methods of Education should not only be adapted to existing differences of temperament; but should be corrective of special and unbalanced development.

3. *What Temperaments need Stimulating to Study?*

From the answer to a previous question it will be seen that, with the exception of the Nervous temperament, there cannot be said to be any difference in this respect; but there is a difference in choosing what means shall be used to stimulate them.

The Lymphatic child is to be stimulated to study by moderation in food. The home arrangements exert more influence in this respect than is usually regarded. Bilious children are to be stimulated to study by encouraging their plays. The teacher should go into the play ground with them, and by vigorous sports awaken the pulsative and vegetative functions, which will re-act on the brain. Sanguine children should be stimulated in the same way, and their studies should be more interspersed with relaxation, in accordance with the demand of their more volatile nature. Thus each peculiar temperament is to be regulated and directed according to its own requirements.

The influence of the seasons upon the respective temperaments, and through them upon the mental life, is a subject of much importance although generally but little regarded.

(4) *What Considerations as to the Different Kinds of Study have Reference to the Several Temperaments?*

The considerations which bear *directly* on the adaptation for a particular kind of study arise chiefly out of the peculiar mental organization of the pupil. These will be discussed in answer to other questions. But the temperaments have an important bearing upon the adaptation of the person to his pursuit or vocation in life; and therefore have an indirect bearing on the choice of studies. Without a knowledge of the mind and the temperamental conditions, in their relation to the practical work of life, young men are often led or put into special vocations by caprice or fancy, or what we may call accidental circumstances; and do not find themselves in the vocation for which their constitution best fits them. But if we observe successful men in various callings, with reference to their temperaments, we find remarkable evidences of the relation between the special vocation and the temperamental condition. Thus, in successful agriculturists, all of the temperaments will be found blended, and the instances of special temperament present a less individualized phase. In these pursuits, in the temperate latitudes, there is, in general, no tendency to a special development of one temperament, but in the torrid and frigid zones, there is an alternative tendency. There is, however, this qualification, that the Nervous element is less prominent than the others, and the Sanguine somewhat more prominent.

Temperamental adaptation to various callings. The description of the Teacher.

The bakers are characteristically predominant in the Lymphatic-Sanguine temperament; for with them a sensitive organization relative to atmosphere, gases, and heat, is requisite. The good cook is proverbially stout and quick-tempered. The butchers are characteristically Sanguine, with a predominance of the organ of Destructiveness. The iron-master and mine-worker will be found usually characterized by the Bilious temperament, with a more metallic and cold, phlegmatic constitution. The scholar and merchant are characteristically of the Nervous temperament.

It is not of course meant that the possession of either of these temperaments is enough to fit for success in the corresponding vocation; but that some degree of it is usually one of the conditions which should be combined with the right mental qualifications, to ensure the best fitness for the work. In each vocation there are special talents, depending upon special order of mental organization.

The importance of the temperaments in reference to the selection of studies consists, therefore, chiefly in their bearing upon the choice of a profession or vocation in life, in reference to which some of the studies to be pursued in the later part of the course should be chosen.

THE QUALIFICATIONS OF THE TEACHER.

A knowledge of the temperaments and of Phrenology affords the means of a definite description of the characteristics of the ordained teacher. He possesses, first, an active sensuous nature, and the elements of temperamental character, in an orderly development; and so far as his temperament departs from such a development, his knowledge of the laws of Physiology, and his constant attention to the subject, enable him to modulate the characteristics of his special phase of temperament, so as to adapt himself to his work in this respect. This sensuous nature and harmonious temperamental character, with its equalized functional activity, is in intimate and sensitive correspondence with the mental life, by reason of a marked development of the sensuous connections of the brain and nervous system, thus bringing his characteristic mental qualities into vivacious and keen sympathy with the juvenile life. The mental qualities which are his characteristic gifts are, a predominant Spiritual organization, first, and above all, Godliness predominating, disposing him to an humble mind,—a leading, restraining force in the region of the Propensities, viz: Cautiousness and Secretiveness,—and a leading development also in the Perceptive faculties, among which Individuality predominates, but also with a marked development of Comparison and Casuality to clearly define principles, and Combinative power necessary to take all the conditions into view at the same time, giving versatility and a ready knowledge of the diversity of influences exercised and combined in the

The gifts of the successful teacher may be acquired.

art of teaching,—and with Aptitude and Spiritual Insight, leading him by their adaptable spirit to feel, forecast, and know the mental state in which he and his pupils are during their labors.

With active senses and this mental and temperamental organization, the varied sensuous impulsiveness and excitability characteristic of children in the schoolroom, will be more or less acutely and vividly realized by the teacher, as they accord with his own mental characteristics in respect to the Social or Animal feelings or Propensities; for all our energies, as a general rule, are resident in the Social group. Hence, in addition to the orderly and regular manifestations of the requirements above mentioned, there will always be a variety of influences which the ordained teacher may skilfully employ, when similar characteristics of the Social and Intellectual faculties exist, to some extent, in his own organization and that of his pupils.

With such gifts, the teacher can modulate and successfully exercise each various element in the temperament and mental character of his pupils, can unite them in one harmonizing spirit, and, by his own special acquirements, and his general and varied powers, can carry them fully to the utmost degree of culture of which they are capable.

In the capacities we have above delineated, are exhibited the order for the acquisition and communication of knowledge, the spiritual power to give the higher scope to the influence and process of education, the sensuous susceptibility to maintain a lively sympathy with the children in his charge, the social feelings to give that sympathy a mental form and realization, and the temperamental adaptation to the task of awakening and controlling attention in pupils of every disposition. It is true that a teacher combining all these gifts is rarely found; but an earnest desire for the possession of these qualities, and industry, assiduity, and self-denial, will enable any teacher of moderate qualifications to approximate to this standard of excellence.

But these gifts, if not already possessed by the gift of the Creator, cannot be attained by effort, without a knowledge of Physiology and Phrenology in their application to education, and a constant labor of self-adaptation.

The actor, although gifted with transcendant poetic and imitative power, needs laborious study and frequent rehearsals, that he may vividly realize in his own mind the leading features of the drama to be presented and the spirit of the part he is to fill, and thus be able to delineate to the audience the feelings and emotions of the character assigned him, with such truthfulness and clearness, that, to them, the actor is merged in the prince or peasant, and they fancy that the real hero is before them. How much more must the teacher, endeavoring to impress upon immature minds practical education, without the attractions of dramatic accessories or illusions, study the minds with which he has to deal, and the peculiar

Teaching an exalted profession—should command best talent and largest compensation.

character of his own powers, that he may infuse himself, as it were, day by day, into their spirits, and instil, without resistance, the instruction he seeks to communicate, and develop into harmony and power the characters which he is aiding to form.

What satisfaction must fill the mind of the teacher who knows and feels in his own experience that he is an efficient instrument of forming the minds of the youth intrusted to his care; and how is this feeling enhanced when he can realize that he is the instrument of the Holy Spirit!

It is obvious from what I have said above, that the profession of the teacher cannot be too highly estimated. It demands, for its highest success and usefulness, a special knowledge and training beyond the scope of the common learning and methodical discipline which it labors to impart. The teacher should know the Human mind, and the bodily conditions upon which the Mind depends.

When the Mind is thus understood, the necessity for the special qualifications of the teacher will be felt, and his gifts and attainments will be appreciated and justly rewarded.

The work of the teacher, then, instead of being a temporary employment which any one may assume for present necessities, and as a stepping stone to ulterior and different vocations, will rise to its true place in the front rank of the professions. It will command the best talent, and be rewarded with the highest remunerations.

The teacher takes hold of Society in its formative elements. When Society knows its own interest in the developing and training process of Education thus definitely understood, and establishes the proper conditions, this vocation will have its own proper precedence.

Until our own time, the world has been without the knowledge necessary for this beginning.

Very respectfully,
Your obedient servant,
JOHN HECKER.

Recapitulation of previous letters.

[*From Mr. Hecker to Mr. Kiddle.*]

HENRY KIDDLE, Esq.,

DEAR SIR: In the preceding letters I have described the general organization of man, with reference to the operations of the Mind, designating the principal organs concerned in the maintenance of life, and indicating what qualities of disposition manifested by the individual are traceable to an origin in the peculiar phase under which the functions of these temperamental systems may be blended in the constitution.

In order to make this clear, with reference to the variable and shifting phases of the temperament during the immature period of childhood, I have delineated the temperamental characteristics, as presented in the more settled and fixed form manifested in adult life.

We have seen that for the purposes of description, four primary temperaments may be enumerated, the brain and the nervous system constituting the organism of the temperament which presides over and is supported by the other three.

The three bodily or physical temperaments—the ærating function of the lungs, the nutritive function of the digestive system, and the secretory functions of the liver—are to the expenditive functions of the Nervous system, what the soil is to the plant which it nourishes.

These three vegetative temperaments combine to supply the organization, which the mind, through the instrumentality of the Nervous temperament, presides over and directs. But although they combine thus, life being the resultant of the associated functions, they struggle for mastery over each other. Great diversities of disposition result from the varying degree in which these elements of life infuse, intersupply, and support each other, within the limits of normal and healthful proportions; and these limits have adaptation to the external circumstances of climate, food, vocation, &c., but undue disparity among these temperamental functions unbalances the constitution.

The attention of the Educator is requisite to promote the equable and harmonious development of the temperament, and he should use the characteristic tone imparted to the mental processes by the temperament as an element in adapting the methods of instruction and discipline. The teacher, himself, is to adapt his own temperamental character to his work, not only through external self-restraint, but, radically, by the power which the mind has, when quickened and guided by the Holy Spirit, to modulate the organic functions of his own physical system as well as of those under his instruction.

The temperaments and mental character delineated as seen in the adult.

Each of the four great elements of temperamental character has its manifestations throughout the system, and may be discerned in the hand, or any part of the body; but they are more readily characterized in popular description, by the form and shape, the complexion, the color of the hair and eyes, and the habitudes of position and motion; and I have described them by these indications.

The immediate duty of the Educator is with the development of the Nervous temperament, and the training of the processes of its chief organ, the brain. The Nervous temperament, in its relation to the other three, may be compared to the E string on the violin. In it appears the highest tension and most delicate susceptibility of the whole instrument; but it depends on the support and harmonious combination of all the others.

The remaining inquiries in your letter relate to the brain, the organ of this temperament, whose processes distinguish Man from the brute creation, and form the chief instrumentality of God's own intercourse with Man.

In describing the cerebral indications of character it will be necessary, as it was in the case of the temperaments, to delineate the more settled and fixed form in which these elements appear in the adult. But it must be constantly borne in mind that childhood manifests the shifting, variable phases of the translucent formative period of growth. The nervous activity dances in the senses and the general sensibility. Every sensuous impression is a novelty. Mental consciousness is undeveloped; purely mental activity waits on the sensuous excitants, and is evanescent and momentary. It is only by slow degrees and long training that the consciousness is developed, and the springs of action, the channels of thought, the prudential restraint, and, lastly, the full physical powers, and the moral susceptibility and impressibility are brought to their due proportion.

While these elements are in the process of formation and the mind is peculiarly susceptible to sensuous influences, the teacher's work is to be done.

As iron will not amalgamate with tin, nor gold with silver, in their natural state, but the skilful artisan, by the insertion of an alloy more fusible than the respective metals to be united, may effect a firm adhesion, so the well-informed teacher, by an intelligent application of Physiological and Phrenological laws, may incorporate his own mental and temperamental dispositions into the minds of his pupils.

In order that the teacher may not be misled by a short-sighted exclusive attention to mere immediate necessities without regard to the ulterior objects of Education, the answers to your questions necessarily involve the whole ground of human development. They require me to touch upon many points which seem foreign to Education as now administered; but which must be taken into consideration, in adjusting the basis of Education in a scientific knowledge of the Mind. The whole field of human character must be included, to elucidate the premises upon which the training processes are to be regulated.

General Principles of Cerebral Form.

I now proceed to consider the questions embraced in that division of your letter of July 27th, which is marked (B).

1. "*What General Principles, (if any) Founded upon External Manifestations of Cerebral Structure may be Adopted as a Guide in Training the Faculties of the Mind?*"

The general principles founded upon external manifestations of Cerebral Structure, which are most important in their bearing upon Education, I will state connectedly as follows, although some of them I have indicated in the previous letters.

1. *Grouping of the Faculties.*—Each hemisphere is composed of three lobes or groups of convolutions, distinguished from each other, both by anatomical evidence of their sensuous connections, and by their contrasted functions. They are the posterior, the anterior, and the upper groups, manifesting, respectively, the passional faculties of the mind, which are termed Propensities, the intelligent faculties, which are termed the Intellect, and the sentient or moral faculties, which I term the Spiritual group.

The characteristic phase of the consciousness varies according as the habitual activities of the mind are centralized in one or another of these groups of faculties.

2. *The Propensities and the Intellect.*—The nerves of sensation from the posterior column of the spinal cord and the thalamus more immediately communicate with the posterior group or Propensities; and here the natural and sensuous forces of the mind have their seat. The Intellect or anterior group is called into action, both by the Propensities and by the nerves of special sense; and the passional desires of the mind, both selfish and social, thus become intelligent. The Intellect, by presenting and individualizing external objects of desire or necessity, re-acts upon and stimulates the Propensities.

The organs of the Intellect and the Propensities meet and over-lay each other in the base of the brain, underlying the Spiritual Faculties above them, and, by the degree of their development, indicating the sensuous force and activity of the mind.

3. *The Spiritual Faculties.*—The upper lobes, lying together along the median line in the crown of the head, have anatomically less immediate and full communication with the sensory and motor tracts than either of the other lobes; and these are the seat of the higher functions of religious impressibility and susceptibility.

4. *The Executive Faculties and the Will.*—The part of the hemispheres having more immediate relation to nutrition and voluntary motion are the convolutions contained in a central transverse core from ear to ear. This part comprises three exterior convolutions upon each side,— *viz:*—Alimentiveness in front of the ears, Destructiveness between or

above the ears, and Combativeness behind the ears,—and also the interior convolutions of the Desire to Live. This part of the brain, especially the convolutions of Destructiveness (which is first in activity) and the Desire to Live, are the convolutions most closely connected with the medulla oblongata, and are the executive faculties, through which physical force of outward manifestation is given to the qualities of other parts.

5. *The Restraining Faculties and the Consciousness:*—The part of the hemispheres having more immediate relation to passional mental restraint, or the control of action and the retention of power, voluntary or involuntary, are the convolutions occupying a region above and behind the ears. The lower or sensuous part of this range comprises the faculties of Cautiousness and Secretiveness; there are two other restraining faculties, which being of a Spiritual nature are properly influenced only through Godliness. The two first named are among the Propensities, and the others lying vertically above them, in the top and back of the head, are the posterior convolutions of the Spiritual group. These four pairs of faculties, I term the Restraining Faculties.

Mental action becomes conscious through these faculties.

6. *Association of the faculties.*—The faculties are not to be regarded as acting individually, each by itself in turn, but in combination or association with each other; and the mental state or act varies according to the faculties which are associated and those which are predominant in the association. The executive faculty, by its greater or less participation in the combination, makes the act an overt one, or a mere mental disposition.

7. *The location of Specific Organs.*—The convolutions in each group, which constitute the organs or instruments of the special faculties, do not have the same invariable position in different heads; but, while they maintain the same general relations of contiguity, have in each head a characteristic development, both in size and in position, whether more forward or backward, upward or downward, and the peculiar form of each head is the result of this combined development.

8. *Sensuousness.*—Convolutions or organs in the lower ranges, in any part of the brain, manifest functions of a more sensuous character than those above them. And an organ developed downward has more sensuousness than the same organ developed upward.

9. *Size and activity.*—Size of the organs is a measure of power; activity, of influence.

10. *Precedence of contiguous faculties.*—The convolutions of each hemisphere have a general correspondence in form with each other, and those pairs of convolutions which lie along the median line, running through the longitudinal centre of the head, tend to possess, by reason of this contiguity, a precedence in activity over the side faculties which are more remote from each other.

11. *The law of life.*—The law of natural life is Necessity, Action, Restraint. It is the primary object of Education to evoke Intelligence to direct the forces which Necessity arouses, bringing into sway the law of Educated life—Necessity, Restraint, Intelligence, Action.

12. *Means of influencing the faculties.*—In Education the mind is approached through the organs of the special senses, and the general physical sensibilities, and these naturally tend to awaken the largest of the sensuous faculties. The way to call into activity a small faculty is to reach it through the largest contiguous faculty.

Some of the foregoing principles will be fully noticed in answer to the specific inquiries you have put. Before proceeding to answer those specific inquiries, I wish to elucidate more fully several of the foregoing principles which are of fundamental importance to a right understanding of what I have to say of the specific faculties.

1. THE GROUPING OF THE FACULTIES.

In a former communication, and by the Phrenologic bust of Washington, it has been explained that the faculties are associated in three groups. You will observe that upon one side of the head, in the Phrenologic bust of Washington, of which views are presented, I have given the general groups and clusters or subdivisions of groups, in which the faculties exist, while upon the other side, is stated the special nomenclature of the individual faculties comprised in each group and cluster. The three pairs of groups and their respective faculties are marked in lettering of different sizes; that employed for the Spiritual faculties being largest, that for the Propensities, smaller, and that for the Intellectual faculties, the smallest of all; corresponding with the gradations, in size and number, of the brain organs. The organs of the Spiritual faculties are the largest in size, but fewest in number, being only seven on each side of the top of the head. Those of the Propensities are smaller, and eleven in number on each side. Those of the Intellectual faculties are the smallest, but are eighteen in number on each side. Each of these groups has a qualitative force peculiar to the faculties of that group, the true nature of which is indicated by the name and order marked upon the bust.

Those organs of the Spiritual group which are in the upper front part of the head, namely, Brotherly-Kindness, Spiritual Insight and Aptitude, constitute a cluster which may be characterized as the Intuitive cluster; while those in the upper back part, namely, Steadfastness, Righteousness, and Hopefulness, constitute what may be characterized as the Meditative cluster. These faculties are brought into a realization of the Truth, and into oneness, by the central faculty of Godliness, if all these exist in their proper order. The faculties of the Intellectual group classify themselves, in the same way, into the Perceptive cluster, the Conceptive cluster, and the Combinative cluster, and the regions occupied by these clusters, respectively, are marked upon the bust. National characteristics largely depend upon the relative predominance of these clusters. Moreover, in any civilized society, men unconsciously classify themselves according to this order, those who have similar predominance in the clusters consorting with each other. When a number of men voluntarily associate together in any one pursuit, there is an organic reason for it, in the similarity of the order of their faculties, or in a complementary relation, by which each supplies what others lack.

By referring to my former letter, in which the names of the faculties are given in their associated order, you will see more clearly the combinations in which they appear upon the bust.

The diversities of mental character, arising from the three-fold division or grouping of the faculties, and from the relative predominance of the groups, may be illustrated and contrasted, by taking, as types of the three classes, the Lawyer, the Theologian, and the Politician. These do not characteristically differ greatly in the circumstances of their development, except in the brief period of professional education, and will, therefore, serve to make more clear the contrast between the groups. Of these classes the Lawyers will afford the most numerous illustrations of the Intellectual or logical mind. Physiologically, they tend to activity in the Intellectual group, rather than in the Social Propensities or the Spiritual Faculties. In mental process, they subject everything to the analysis of the faculties in the Intellect, particularly in the Perceptive cluster. It is their forte to perceive and discriminate clearly, and to command the resources of language, both in speaking and in writing; but all this they may possess, without that executive force and administrative ability, which rest in the predominance of the Animal and Social Propensities, led by Destructiveness, (or Executiveness,) one of the faculties of that group. Moreover, they may possess either or both of those characteristics, without that quality which is termed the Judicial mind, which results from the predominance of the Spiritual group, led by Righteousness and Steadfastness, or as these faculties, in their natural development, are called, Conscientiousness and Firmness.

The Theologian will serve as the type of the class in whom the Spiritual Faculties are predominant. He seeks, not merely to discern truth by means of its outward, objective forms and proofs, through the Perceptive Faculties, but strives rather to receive and realize it in his inward consciousness; and this is to be done by the direct influence of the Holy Spirit upon the soul in its consciousness in the Spiritual Faculties. He claims that by this Spirit he has been called to minister to men in holy things. There are many Theologians who reason only intellectually like the lawyer; but, though Spiritual Truths may be analyzed and expounded in this way, the disquisition which the Intellect gives cannot directly, though it may indirectly, induce a Spiritual realization of the Truth in the consciousness of the hearer. The habit of using certain sounds and formulas of words for the expression of the activity of certain faculties, gives only the objective means of awakening and exercising those faculties in another person. In order to bring the truth directly to the realization of the spiritual consciousness of the hearer, it is necessary that what is expressed should be first realized in the spiritual consciousness of the speaker. If it is so possessed by him through the instrumentality of the Holy Spirit, the communication of the truth by him will be with power. This spirit in him gives the inward or subjective condition for religious teaching. The one without the other is the form of Godliness without the power thereof.

The Politician is successful in his peculiar calling, because of the predominance in him of the Social and Animal Propensities. It is in these that the forces reside, and if these organs are proportionately largest in bulk, there exists a great activity, as well as force, in these faculties. If, instead of being drawn forward into the Intellect, or upward into the Spiritual group, the forces are exercised in the Propensities, and are sufficiently restrained by Cautiousness and Secretiveness, the man will possess that energy, pugnacity, social influence, and tenacity which enable him to lead other men in public affairs. Men in successful political life will be found, by the observer, to possess breadth of head in the region of the faculties of Secretiveness, Cautiousness, Destructiveness, and Combativeness.

A degree of the same development in the Theologian, accompanying the predominance of the Spiritual Faculties, constitute the good organizer, and tends to make a leading ecclesiastic.

These generic distinctions between men, arising out of different predominance in the groups, are of prime importance, and are to be kept in view at every step in the discussion of mental science, or of the history of opinion.

LANGUAGE.—The mental differences resulting from this difference of organization affect the use of Language, which is the representative form of thought.

Language is an expression of the public mind. No man can give a precise and exclusive definition of words, because it is the result of public usage which is multiform and diverse. Language cannot be accurately and precisely understood except with reference to the diversities of character dependent on the activity being centralized in different groups.

Men understand language, and use it, with different significations, according to their organization in this respect; and the very word with which we designate a mental phenomenon may represent an essentially different fact, according as the one group or the other predominates in the person speaking or hearing it.

A description of the character and functions of the Spiritual Faculties especially cannot be rightly understood, without attention to this relation between language and mental organization.

The word *Love*, as the expression of a mental phenomenon or state, in a mind in which the Spiritual group predominates and has been awakened, designates a different state or affection from that designated by the same term, when used by a man in whom the Spiritual Faculties are dormant and the Propensities predominate. The former is a principle not inherited, but the reflection of the Divine Nature, wholly unselfish, having for its objects, primarily, the Creator, and next, the welfare of one's fellow-creatures. This is the love which the Gospel teaches, which was manifested in Christ, and of which the Apostle speaks, when he says— "Let us love one another : for love is of God : and every one that loveth is born of God and knoweth God. He that loveth not, knoweth not God : for God is love." The other affection or state designated by the same term, love, is the activity of Adhesiveness, or some other of those social affections, or even mere Animal desires, which are among the Propensities, the love of husband and wife, love of offspring and children, love of the race, home, love of self, love of friends. All these, beside other Propensities, are implanted by the Creator, and have for their object the continuance of the race, upon earth, and the social harmony necessary for that object.

Faith is the conscious possession and realization of God by His Holy Spirit, when the consciousness of the soul is centralized in the Spiritual Faculties. In this spiritual sense. it is a generic term for the manifestation of this group. It is not to be confounded with Intellectual belief or logical conviction, which is the assent of the Conceptive and Combinative Faculties to the deductions from outward and perceptive facts.

Patience, in the truest sense of the term, is the Spirit of Godliness, Brotherly Kindness, Steadfastness, Righteousness, and Hopefulness, manifested in long suffering. In the worldly sense, it is Restraint through the faculties of the Propensities—Cautiousness and Secretiveness—for interest's sake to bear and forbear in order to get the advantage.

Joy, in its Spiritual or Christian meaning, is the realizing sense of the indwelling of God's Spirit—joy in the Holy Ghost. In the worldly sense, it is the immediate gratification of the desires of the Propensities.

Many other words used to designate mental states, are equally ambiguous.

2. THE PROPENSITIES AND THE INTELLECT.

These two groups, constituting together the whole lower part of the hemispheres, are the faculties through which man in the natural state manifests his life.

All the mental operations of these groups are, in their nature, physical; and upon their quality and proper exercise the whole physical forces of the system, and the natural mental disposition depend, for control and direction.

All of the Propensities but two tend to action, each following its own law, and in the combined activity their united force follows the characteristic law of the group; *viz:* Desires.

The desires are all dependent upon sensuous conditions. Those of the most sensuous character, which have their organs at the base of the brain are of a more gross nature, their mental character being blended with the physical or bodily functions. These are called, distinctively, Appetites.

The Desires, that is the activities of the Propensities, when intensified beyond the ordinary degree, are termed the passions.

The Necessities of Life reside in the Propensities.

The Intellectual group of faculties are not emotional like the Propensities but cognitive. In them resides not the force but the direction of life.

Intelligence without Desire is inactive.

Our poor houses are full of people with predominant Intellectual faculties, who by reason of deficient force or restraint in the Propensities (in the absence of Spiritual guidance) are improvident.

On the other hand our prisons are full of persons with predominant Propensities, but deficient in restraint and Spiritual guidance.

The Desires of the Propensities invoke the activity of the Intellect giving intelligent selfishness. Cunning is the lower form of such intelligence, where the sensuous faculties at the base of the brain are most active. Prudence is its higher form, where the restraining faculties and the conceptive and combinative reason unite in guiding the action, under some influence of the higher moral sentiments.

The Spiritual faculties are not brought into their true life by these sensuous groups. Man is lost without special Divine guidance.

3. THE SPIRITUAL FACULTIES.

By nature, the Spiritual group is not predominant in activity. Unless energized by the power of the Holy Spirit, its faculties are passive; for, although they have a certain natural exercise, it is subordinate to the Propensities or the sensuous Intellect, and does not amount to an activity which rules them.

The true function of these faculties is reflective, in the strictest sense; and the necessary condition for this function of impressibility and susceptibility, is the superseding of the sensuous activity of the predominant faculties in the Propensities and the Intellect, by the meek and humble activity of this whole group. The peculiar qualitative character, only, of the manifestation depends on the size and order of development of the organs, and the temperamental conditions.

Reflection defined. Manifestation of the Truth by the Spiritual faculties.

Metaphysicians, not recognizing the reality of a Divine influence upon the hearts of men, have defined reflection, as the process by which the mind turns itself back upon itself and its own consciousness; and have asserted, that every thing exists previous to reflection, in the consciousness—this function of the mind being to consciousness what the microscope and the telescope are for the natural sight—not making the objects, but illuminating them, and discovering to us their character and their laws.

But the mind does not only turn itself back upon itself in reflection; it may turn itself toward God, so to speak, and receive His influence by reflection. The Scriptures and the religious history of man, and our own experience, alike teach us, and the science of mind establishes, that the Truth, which is single, entire, and absolute, is reflected in the mind of man, by the instrumentality of the Holy Ghost, in the Divine order, *viz:* Godliness, Brotherly-Kindness, Steadfastness, Righteousness, Hopefulness, Spiritual-Insight, and Aptitude.

It is this reflection of the Truth, in its oneness, which is the true function of these seven-fold Spiritual gifts.

They receive the Truth passively, as it were, and become consciously aware of its power; and, according to their order, they shed it within the consciousness, and upon all the faculties, and upon mankind around.

This reflection varies in accordance with the organic order of these faculties in the individual, and the resulting gifts proceed in accordance with the order of development. The reception of the influence of the Holy Spirit does not necessarily or immediately modify the order of development of the Spiritual Faculties among themselves, or as compared with the others; but it does give them, as a group, immediate commanding *influence* over the other two groups, the Intellectual Faculties and the Social Propensities,

When all the Spiritual Faculties are developed, in their true order, as above stated, and, by the power of the Holy Spirit, thus lead the faculties of the other groups, the Propensities are subordinated by the power of the Spiritual nature, and their centralized will and worldly Spirit are superseded; while their various faculties, particularly Destructiveness, Cautiousness, and Secretiveness, are brought into requisition to execute the Spirit's behests, and the Intellect also is called to answer its demands, both in synthesis and analysis. Thus God works in us, "both to will and to do of His good pleasure."

It is not, however, left to individual man to manifest the Truth. God, through Christ, manifested the Truth in its perfect order in human organization. Christ's conversation with his disciples before He suffered, and His prayer for them, and for those who, through them should also believe, teach us that He came to manifest the eternal life, that is, He came in order that men might know God; that the Spiritual life and keeping of His followers depend upon their being united in Love, as he organized them; and that their sanctification depends on the Truth, which He asked the Father to send them; that, thus organized and guided, they were sent into the world, though they were not of the world, to the same end that the Father sent the Son into the world; and that it should be through their Word that others should believe on Him; and that all disciples must be united in love, in the method of His appointment, in order that the world may believe on Him. Thus, by the Church, which Christ founded, associating with Himself twelve persons, the unity of the Truth is to be manifested in its harmony and completeness, as it never could be through individuals with their diverse organizations and remaining sinful dispositions, or through more numerous bodies.

NOMENCLATURE.—The method by which the location of the organs has been usually defined by Phrenologists is to indicate the region occupied by each one, when it is predominant, and not specially modified by other faculties or by the unequal development of its own parts; and the method by which the faculties have been named by Phrenologists is to take the extreme manifestations of each, when not specially qualified by others, to characterize its quality.

In entering upon the subject of individual faculties, it is necessary to recognize that Dr. Gall, in his own mental organization, was disposed to conceive a principle upon the suggestion of a fact or phenomenon, and then sought for other facts to support the principle to its fullest extent. Dr. Spurzheim, by his organization, was disposed to perception, and hence, by more special observations, was led to narrower deductions. He saw the necessity for a system based in a constructive order, upon the perceptive facts observed in general and in detail. In the main, his more specific investigation substantiated and gave point to the principles conceived by Dr. Gall. But his nomenclature and classification were made without sufficiently discriminating the basis of Phrenology in the localized functions of the faculties *in groups*.

Dr. Spurzheim carried the science to a much more arbitrary presentation than it had before received, no man since his time having contributed so much to its progress; but his philosophic classification is inadequate. Dr. Gall's philosophy in respect to the form of the head, was sound, but the facts were too arbitrarily individualized and presented by him, to give the subject its proper place in the minds of sensitively intelligent men, as a science. But in the infancy of the science such a nomenclature was essential for the establishment of Phrenological truth as he conceived it. To illustrate this, Dr. Gall first characterized a particular faculty as the organ of murder, because he found it largely developed in every murderer's head; but Dr. Spurzheim was obliged to modify this, and called it the organ of Destructiveness, because of its prominence in many of a quiet and peaceable disposition; by Phrenologists of the present day it is regarded as simply an executive faculty, and will, I doubt not, ere long be recognized as the organ of "Executiveness;"—and thus, as the science advances, the whole nomenclature given to the passions will be modified on an intelligent basis, presenting the mind in a more humane aspect.

In defining the qualitative character of the faculties, and delineating the position of the organs, it is of the utmost importance to regard the variations which are caused by the influence of the group which may be predominant,—by the force of large Propensities, by the modifying influence of cultivated Intellect, by the influence of the natural Spiritual Faculties,—and also by the consorting of the special faculty in question with faculties contiguous to it, whether in the same group with itself or not. With the exception of the awakened Spiritual Faculties, which illuminate and imbue the whole being, each of these conditions affects the position and general shape of the organ and the qualitative character of the faculty, and often, indeed, the whole shape of the head, and must be taken into consideration, in a correct analysis of the faculties in any given mind.

I have adopted the names given by Dr. Spurzheim to all the faculties, excepting those in the Spiritual group. In delineating the faculties of this group it must be observed that we are contemplating, not man as an animal, merely, perverted and lost, but also man redeemed, and restored to Spiritual life in the image of his Creator. The reader of Drs. Gall and Spurzheim's delineation of the faculties will see that, even in regard to these Spiritual Faculties, which the latter terms Sentiments, they look among the animals for the proofs of their existence, and when they describe them as manifested in man, it is chiefly in their natural and low state. Thus regarded, the nomenclature which Dr. Spurzheim adopted is not inapt; but the true activity of these faculties as they should be awakened under Christianity, requires names of deeper significance than mere Moral Sentiments.

GODLINESS.—This faculty was first designated by Dr. Gall, the founder of Phrenology, as the organ of Theosophy, or the organ of God and Religion; and in giving it this character he was right. But Dr. Spurzheim, who treated Phrenology by an Intellectual analysis, denied that man can know God; and stated that this faculty is only a sentiment, and is blind. He accordingly termed it Reverence.

If we ignore the true life of the Spiritual Faculties, and regard man only in his fallen estate, and unregenerated, this would be a just description of the faculty. Dr. Spurzheim did not sufficiently regard the facts of the religious nature and history of man. God does make Himself known to men, by His Holy Spirit, who has direct, immediate intercourse with them through this faculty, and through it illumines, first, the Spiritual nature, making it HOLY, and thus illumines the whole mind, giving it a godly character, I have adopted, to designate this faculty, the term Godliness, which is that used in the Holy Scriptures.

The natural tone and manifestation of the faculty of Reverence or Godliness, is humility, and it is to this receptive frame of mind, which the Scripture characterizes as the humble and contrite heart, that God promises His presence and grace. But the mind is by nature indisposed to receive the Spirit of God, because the Propensities predominate, and, by calling the Perceptive Faculties, which are sensuous, into their service, rule the Intellect and overpower the Spiritual Faculties. This is the carnal mind, which is enmity against God. When the Propensities are strong, therefore, they intercept the exercise of Godliness; and for this reason, the passions must be subjugated and held in control, in order that the mind may be in this meek, receptive state.

The faculty of Godliness is, as it were, the eye of the Soul. Its special function is to receive the Holy Spirit, as the eye receives natural light. When the mind examines its own consciousness in comparison with the indwelling consciousness of the Spirit of God manifesting His influence in the heart, a self-evident conviction of our own shortcomings and sinful state arises. It is apparent that while God is good to all, the just and unjust, and manifests His Divine love to all mankind, we, however well disposed, are imbued with selfish motives, and do not fulfil this law of love. This self-examination and consciousness of sin, in contrast with the Spirit of Godliness, give the opportunity for growth in humility and grace, and open the mind more and more to the renewing power of the Holy Spirit.

This fitness to receive the grace of God, and to become transformed by Him, is increased and is afforded its true conditions, when the faculty of Brotherly-Kindness is next predominant, in accordance with the order which Christ established, He declaring, that when two of His disciples agreed touching what they asked, they should receive, and that where two or three were together, He would be in the midst of them.

BROTHERLY-KINDNESS.—Dr. Gall and Dr. Spurzheim designated this faculty Benevolence. The former was somewhat at a loss to distinguish it from Conscience and the Moral sense. Dr. Spurzheim's observation of this faculty seems to have been more specific, and he describes it as different from the moral sense, and as a fundamental power, producing mildness and goodness and a long catalogue of modified actions variously styled benignity, clemency, mercifulness, compassion, kindness, humanity and cordiality.

I place it in next order to Godliness, because this is the order of the Spirit of the Holy Ghost, which makes love to man the second great command. Godliness is the regulating force of Brotherly-Kindness.

The Faculty of Brotherly-Kindness.

This is the foremost of the sympathies. When awakened by the Holy Spirit, it feels the obligation and the duty to bear the sufferings of one's fellow beings; and brings into requisition all the faculties of the mind to carry out this necessity. It is in this faculty that the virtue of Christian helpfulness finds its practical exercise, in the same sympathy which led our Saviour to suffer on the Cross for mankind. It was through this faculty in the Apostles, that Christ's promise to give the power of curing diseases to those who had faith, was fulfilled in them. The ordinary benevolence of men has relation chiefly to the outward and bodily wants of those in distress. The full and Christian activity of this faculty, when awakened by the Holy Ghost, has regard to man at large, without respect of persons, although its more specific exercise regards those fellow-creatures, who are dependent on the individual, or immediately around him. Its object is not merely the bodily and temporal welfare of men, for its own sake alone, but it cares more especially for their spiritual life, and for their bodily welfare as the condition of inward life.

The exercise of this faculty is one of the chief conditions of maintaining the presence of the Spirit of God in our hearts. It is, by its position, the faculty most powerful to support that of Godliness, and, by co-operation, to increase the activity and influence of the latter, which should always lead it. The young ruler who came to Christ, had fulfilled all righteousness, but lacked yet the witness of eternal life in his heart; and our Saviour, who knew what was in man, directed him to abandon himself and all that he had to the exercise of this faculty, by selling all and giving to the poor, through which he might have attained the power of Godliness. The teachings of scripture, the principles of mental science, and the experience of the Christian alike attest that self-denying sympathy and care for the wants of the needy and suffering, especially those who are not naturally congenial to us by character or association, and whom the Propensities would forbid us to serve, is the first external or objective method of growth in grace, being the exercise of the faculties next in order after those of Godliness.

The organs of these two faculties, Godliness and Brotherly-Kindness, are located contiguous to each other, in the central portion of the head. The right exercise of each depends on the exercise of the other. "If a man say he love God, but love not his brother, he is a liar, and the truth is not in him."

Ethical teachers are accustomed to put Conscientiousness, or Righteousness, before Brotherly-Kindness, in importance. But our Lord Jesus Christ teaches us that supreme Love to God, and a Love to our neighbor equal to that for ourself, are the first two principles, upon which all others depend, and are equal in strength in God's sight to all the desires of the Propensities. The leading principle is Love, and Righteousness is to be inspired by it and to modulate its action: Righteousness is the breastplate; its function is to guard and restrain.

The Faculty of Steadfastness.

STEADFASTNESS.—This is the faculty which Phrenologists, not sufficiently recognizing its higher relations with Spiritual qualities, have designated as Firmness. It has no direct relation to external objects; but its function is to add its own positive quality to the manifestations of the other faculties. Thus, in combination with Self-Esteem, the contiguous faculty of the Propensities, it increases the strength and individuality of the personality; in combination with the social affections, it assists in giving constancy to those affections: with the executive and administrative faculties, it tends to stability; with the Intellectual Faculties, it gives steadiness and permanence. It is often said that, if this faculty is deficient, the person is yielding, and pliant, and subject to follow the wills of other persons; but this is more often otherwise; for though it be deficient, the Faculties of Self-Esteem, Inhabitiveness, Philoprogenitiveness, Approbativeness, and others of the Social Propensities, still may, and commonly do, give persistency. If it is too predominant, and unregulated, it necessarily shows a strong individuality, and it overrules, or rather, holds back other faculties, and results in obstinacy and stubbornness.

Dr. Spurheim defines this faculty, as he does others of the Spiritual group, as a peculiar natural sentiment; and in delineating its influence upon the character, he does not go beyond the scope which it has in the natural and sinful state of man. This view necessarily resulted from his method, in which as I have explained, he took man as found in his natural fallen state, and consequently delineated every thing in the condition of the predominance of the Propensities and the Intellect.

When, however, this faculty is awakened to Spiritual life, it finds its true function in its relation to the central faculty of Godliness, next to which it lies. That staidness which is inspired and regulated by the love of God, is the true quality of this faculty, and gives to the whole mind a nobler character than any mere sentiment of firmness. This quality the Scriptures designate Steadfastness, and I have adopted that term. Even though Steadfastness predominate over the Propensities, as it ought always to do, yet as long as it stands in its proper order toward the other Spiritual Faculties of Godliness and Brotherly-Kindness, which should precede it in order, it is incapable of that abuse which we have indicated as obstinacy. If it be ruled by love to God and man, and combined with Righteousness, which is next contiguous to it, it gives the highest inflexibility of heroism and martyrdom, without stubbornness or intractability. In the knowledge of God, the mind finds the ground and rock of Steadfastness, always looking at the eternal and immutable, and, in feeling, resting upon them. Thus Asa cried in his prayer for help against the invasion, " We rest on Thee, and in Thy name we go against this multitude." The Psalmist gives constant expression to this fixedness of heart, and points to Godliness as the ground of it, when he says,

"I have set the Lord always before me; because he is at my right hand, I shall not be moved," and again, the righteous "shall not be afraid of evil tidings; his heart is fixed, trusting in the Lord." And the Apostle delineates this virtue of Steadfastness as related to the right activity of the Spiritual Faculties, calling for the whole armor of God, that we "may be able to withstand in the evil day, and having done all, to stand."

This is the first of the *Restraining Faculties* of the Spiritual group. Through it, in connection with the other restraining faculty, Righteousness, comes control, as well as the retention and continuity of power.

RIGHTEOUSNESS.—This is the faculty which, by Dr. Spurzheim, is called Conscientiousness. It is the moral sense, the sense of right and wrong and of moral obligation. In the natural state of man this faculty, like the other Spiritual Faculties, is subordinate to those of the other groups; and metaphysicians, analyzing their own consciousness, or studying the consciousness of men at large, have, of course, been unable to agree upon the nature of this feeling. Some have asserted that the moral sense arises from self-love, that is to say, that the ultimate test of right and wrong is, what is, in the highest sense, for our own interest; others, that the love of praise is the source of this feeling; others still, that it is a deduction by reflection, from benevolence and sympathy; others, again, having perhaps higher mental organizations in view, have traced it to a sense of the fitness of things, or the hope of eternal welfare.

Now, in point of fact, in a man in whom the Spiritual Faculties are not predominantly active, if the faculty of Approbativeness among the Propensities leads the mind, Conscience or the moral sense will be subordinated in activity to that; and it is a just description of this faculty in such a mind, to say that it depends upon the love of praise. Again, in a mind in which Cautiousness and Secretiveness predominate, especially if Hope is also large, the judgments of Conscience will be, as some philosophers have declared them to be, based on utility and the fear of evil. In the same way, all other theories of Conscience will be explained, if the combinations of other faculties with it be considered, and whether the one or the other predominate in the combination. These facts elucidate both the diversities of the action of Conscience in different minds, and the contradictory theories which philosophers have formed in regard to it.

There is a special sense of right and wrong resting in the Conscience, in the limited conditions of moral or spiritual life to which man is by nature disposed, and hence many ethical philosophers, ignoring the spiritual judgment, have by a sort of eclecticism made Conscience dependent upon fortuitous circumstances. This is as far as philosophy seems able to go.

The Faculty of Righteousness. Hope, or. Hopefulness.

When the Spiritual Faculties are awakened, the moral sense is no longer a sentiment, led by and depending on analytic or selfish faculties; but its true individual character and relative order appears. It is illumined by the Holy Spirit, and controls, stimulates, and reproves all the activities of the mind. This is Holiness, without which no man shall see God. In its fulness, it was manifested by our Saviour Jesus Christ alone. If we seek for Truth in its singleness and entirety, and, in the humble spirit, strive to receive it, hungering and thirsting after it in the way of Godliness, we shall all receive the same Truth alike, as it was manifested in Christ Jesus, as He declares Himself, "I am the Way, the Truth, and the Life;" and when we understand the diversities of mental organization we shall all unite in the absolute Truth, although diversities of manifestation will continue. When Conscience is thus enlightened, being, as the Scripture terms it, Righteousness, it bears witness in the heart that one is the child of God.

If Steadfastness be insufficient, and the practical life, therefore not held in conformity to Righteousness, the latter faculty acts rather by reproof and self-condemnation, than as a guide and a source of confidence.

The Apostle indicates the relation of this faculty of Spiritual knowledge and judgment to those of Godliness and Brotherly-Kindness, which, in the true order, lead this, when he says, "I pray that your *Love* may abound yet more and more in knowledge and in all judgment."

These two last described faculties, Steadfastness and Righteousness, are the *Restraining Faculties* of the Spiritual group. In this respect, their influence has some analogy to that of Cautiousness and Secretiveness among the Propensities. Through Steadfastness and Righteousness, come that weight of judgment, that abiding in the Truth, that soberness and vigilance, which are proper and right in all things. These qualities are peculiar to Christian virtue, because they derive their inspiration, not from the Propensities, but from the knowledge and love of God, as manifested by Christ Jesus, and shed abroad in the heart. A man in whom these qualities are strongly marked, and who also possesses natural force and intelligence, must lead others, because others of less strength in these faculties will lean upon and follow him.

Steadfastness, when in combination with Righteousness, and with Cautiousness and Secretiveness, and led by the predominance of Godliness and Brotherly-Kindness, gives that prevenient grace which is so little understood in the Christian life. It was through this combination, under circumstances of overwhelming necessity in the surrounding condition of men, that the gift of prophecy, by the grace of God, resulted.

HOPEFULNESS.—Dr. Gall considered Hope as belonging to, or a part of the function of every faculty. Dr. Spurzheim criticized him in this respect, saying that he confounded this peculiar feeling with desire or want;

The Faculty of Hopefulness.

and Dr. Spurzheim describes buoyancy and elation of spirit, and the confident expectation of success in whatever the other faculties desire, as the function of Hope; adding, however, that this sentiment is not confined to the business of this life; but inspires hopes of a future state, and belief in the immortality of the soul. The Scripture delineates Hope among the noblest faculties of the soul, and shows its essential importance; and Dr. Spurzheim was doubtless right in distinguishing it from the anticipatory affections of the other faculties.

But Dr. Gall was also right to this extent, that a great part of what men term Hope is merely a vivacity in such affections or activities of other faculties, arising in part from a peculiar lively sensuous vitality in a special order of the Propensities, and in part from the temperamental conditions, and is not the activity of this organ of Hopefulness.

In its natural manifestations, Hope is generally subordinated to the faculties below it, and not to those above it; and takes its character, not from the Truth, as is the case when Hope is centered in God, but from the desires of the Propensities. Thus, in connection with large Acquisitiveness, it gives the hope of success in business; with large Cautiousness, the hope of safety in danger; with large Approbativeness, the hope of fame.

In all these manifestations, Hope is notoriously illusory; comforting and encouraging for the immediate present, while its anticipations are constantly disappointed. If the faculty be large, advantages are magnified, and obstacles forgotten, the person procrastinates, and unless Cautiousness and the moral sense are strong, he will be lavish in promises, which will go unperformed.

When this faculty is awakened, and stands in its true order, having its activity predominantly in combination with Godliness, the Truth inspires and guides it, and it becomes characteristically a sober and just anticipation of the future. Instead of being led by the outward lively sensuous nature and the desires of the Propensities, it then leads and inspires them, they being kept, however, within the just limits marked by the influence of the Holy Ghost manifested through the Faculties of Godliness, Brotherly-Kindness, Steadfastness and Righteousness. Then is given that fulness of Hope, which marks the spiritual state of the Christian. In its proper order among the other faculties, and having its right exercise, it characterizes the whole phase of the mind, and illumines and draws forth the efforts of all the other faculties in their due order, in this characteristic, answering very nearly to Dr. Gall's view of it. Hence to give a true description of this faculty, in its spiritual activity, as it is delineated in Sripture, we may designate it as Hopefulness. This term corresponds to Dr. Gall's characterization of this quality of the mind, and includes Dr. Spurzheim's view.

The quality of vivacity and sensuous activity which characterizes youth, and is often spoken of as Hopefulness, we understand to be, not so much a mental state or faculty, as a sensuous pre-occupancy of the mind in external life which precludes or supersedes despondency, and alike precludes or supersedes, to a degree, the activity of the other higher faculties. The Faculty of Hopefulness, when awakened, is no longer a specious sentiment, capable of deluding the mind, but is "the anchor of the soul, both sure and steadfast." The Apostle delineates the exercise of this faculty, in its dependence upon Godliness, when he says that "Hope maketh not ashamed, because the love of God is shed abroad in our hearts by the Holy Ghost, which is given unto us." But, on the other hand, when the Propensities rule the mind, this faculty depends on them, with their sensuous relations to the external world, for its activity, and lends its character, as it were, to them, increasing their delusions; and as the communication of its organs with those of the Propensities is through the organs of Approbativeness, contiguous to which Hope lies, it is in such a mind most frequently excited under conditions which predispose the mind to shame.

SPIRITUAL INSIGHT.—This faculty Dr. Gall observed to be prominent in all persons he met with, who were prone to believe in apparitions and supernatural marvels. Dr. Spurzheim, who was first inclined to term it Supernaturality, afterward designated it Marvelousness, because, as he said, it may be excited both by natural and supernatural events, and in every case fills the mind with amazement and surprize. By Dr. Combe, it was termed Wonder.

In thus designating it, they have characterized it by a sensuous and an extreme and special manifestation, and have not recognized its proper and most useful practical exercise, when combined with other Spiritual Faculties, which is the recognizing and being impressed sympathetically by the spirits of other persons. Phrenologists have been led to conjecture and assert, that there is a faculty of the mind which gives an instinctive knowledge of character, by its power to recognize and sympathize with the natural expression of the feelings of other persons which is marked in their countenance and in the whole person. To the unusual development of this faculty, in such men as Bacon, Shakespeare, and Scott, has been attributed their deep insight into human nature. There is without doubt, a constitution of mind which possesses this power, and its qualitative character is as various as the organizations are various.

Careful observations will show that these functions of Wonder, Marvelousness, and Discernment, are different manifestations of one and the same faculty. This is the faculty which gives the successful public speaker his sympathetic and intuitive possession and understanding of the minds of his hearers; it is also the faculty which adds enthusiasm to the motives of men socially united.

By this faculty, the soul, when illumined by the Holy Spirit, receives the influence of other minds, and enters into and possesses them by the direct reflex action of the Spirit of God.

The Faculty of Spiritual Insight.

Some writers on Psychology have recognized and described the singular fact which sometimes occurs in the experience of thoughtful persons, that one seems to be made vaguely conscious of the bodily presence or even of the thoughts of another, without any apparent external suggestion of the approach or of the mental state, as the case may be.

This is particularly observed in the case of intimate friends and companions. The mind of one has a premonition or presentiment of the most unexpected meetings, or the words of one seem to be the expression of the very thought passing through the mind of the other, under circumstances which will not account for this sympathy or identity of thought by the existence of any common external cause or by association of ideas. Other writers have denied the possibility of any such faculty, explaining the alleged instances as accidental coincidences of electric laws.

The phenomena referred to depend, usually, on the exercise of the faculty of Spiritual Insight; but its exercise is unconscious unless by a special knowledge of the faculties brought under attention, or unless the Spiritual Faculties are under Divine influence.

It is through this faculty, when awakened by the Spirit of God, that we have an inward conscious knowledge of our own hearts as well as of others. The discerning of Spirits, which is one of the gifts of God mentioned by the Apostle Paul, is through this faculty. When this faculty is large, and ruled by large Perceptive Faculties, if excited by the influence of the Spirit of another, or any subject in which the mind centralizes itself, it gives the vision of apparitions, and the perceptive powers being overshadowed by imagination, the man feels, sees, or hears what has no outward objective existence, but is merely the result of the activity of the faculty of Spiritual Insight, in combination with predominant Perceptive faculties. Other manifestations come from the predominance of the Conceptive or Combinative clusters respectively. In these special manifestations, Marvel or Wonder is the result of the combined activity; but the proper function of this Spiritual Faculty is the discernment of that which thus influences the individual. When the Spiritual Faculties are awakened, and, in their proper order, lead the mind, this faculty has its most useful and practical exercise in this discernment; and its proper scriptural designation is, therefore, Spiritual Insight. It should be by this faculty, acting in its proper order, and illuminated and guided by the Holy Spirit, that those who offer themselves to the ministry of Christ, professing to be called by the Spirit of God, should be tried, to see whether they be of God. It also gives the realizing sense of everlasting life, and that Spiritual unity in which the hearts of Christians blend in true worship.

The character of this faculty, and the teaching of the Scripture, lead us to believe, that when the kingdom of God comes, and His will is done on earth as in heaven, the Church will be one, all its members being one with another in Christ, as Christ was one with the Apostles and with the Father, according to His prayer on behalf not only of the Apostles alone, but also of all who should believe in Him through their word.

Spiritual Insight. Imitation, or Aptitude.

When the fulness of the Spirit is manifested in men through the order taught by the Scriptures, this unity will result; and it is only from this manifestation of the Divine Spirit, that the gifts and graces which are promised to the Church, will come.

In its natural state, unenlightened and blind, this faculty, being subordinate to the outward sensuous Propensities, serves them, and is led by them into the errors to which they are prone. In this state, it tends to superstition and credulity; and, with Aptitude, or Imitation, it disposes to panic when men are influenced by a common danger. In men in whom vicious activities of the Propensities rule the mind, this faculty, if large, gives readiness in discerning who are susceptible to their evil influence, and facilitates the power to seduce others into sin.

From this brief description of this faculty, it will be seen that its right exercise is very important to the teacher. He is called on to guide the mind of the child, and sympathetic relations between him and his pupil are essential to his success. He cannot have the same natural and instinctive sympathy which the parent has; but Spiritual Insight, if it be developed in the teacher in its proper order, and awakened, enables and disposes him to enter into the mind of the child, discerning his mental processes; and if Steadfastness and Righteousness are large, so that his judgment is good, he then intuitively understands and appreciates the misapprehensions and the motives of his scholar. His explanations and admonitions will not be wasted where the scholar does not need them; but he feels the root of the difficulty with which he has to deal, and all his instructions are directed to the very point at which they are needed.

APTITUDE.—This is the faculty usually called Imitation. Dr. Gall was first led to recognize this organ, by observing its prominence in a friend having remarkable powers of mimicry; and subsequent observation of other persons possessing similar powers led him and Dr Spurzheim to characterize it as the faculty of Imitation, and to refer its importance largely to its dramatic function.

The ability of the dramatist or actor consists of his mental impressions in this faculty, and his skill in giving them outward form in language or demeanor.

This faculty, in its individual activity, or where it is predominant in connection with Constructiveness or Mirthfulness, and is led by the Propensities, tends to these dramatic manifestations; but its proper manifestations are of a more general character, and of greater importance. It is this faculty, especially when combined in activity with Approbativeness, that makes men conform to each other in society, and reduces individual idiosyncracies so as to produce a degree of harmony, in following the general, social standard. Fashion and the uniformity of manners and customs among any given community, depend upon this faculty.

The Faculty of Aptitude. Spiritual Insight and Aptitude of the Teacher.

In children, this faculty is very much exercised in connection with the Sensuous faculties, which lead it. It gives children the disposition to do as they see others do, which is a powerful instrumentality in education.

If the Propensities are predominant over the Spiritual group, the individual will be impressible and comformable, easily catching the spirit of his companions whatever their influence may be, and readily assuming the same characteristics that mark their motives and conduct. This faculty increases the susceptibility to be enticed by others into sinful indulgence.

But if the Holy Spirit has been received by the soul, and love to God and man rules the mind, through the predominance of the Spiritual group in the true order, this faculty of Aptitude tends to bring the individual into harmony with the Divine Spirit. Then the other Spiritual Faculties Godliness, Brotherly-Kindness, Steadfastness, Righteousness, Hopefulness, and Spiritual Insight, leading this, the person, instead of taking for his imitation the opinions and the conduct of others about him, makes his standard the Divine law of love. This contrast the Apostle points out, when he exhorts not to be "conformed to this world," but to be " transformed by the renewing of your mind." This faculty, Aptitude, gives the characteristic spirit of a disciple or follower. Spiritual Insight disposes us to enter into the example of Christ, and makes it a living power in the heart; and Aptitude disposes us, by sympathetic influence, to manifest the spirit in our lives.

These two faculties, together with Brotherly-Kindness, constitute the cluster of Intuitive Faculties, which give the impressible and teachable character to the soul; and when Christ said, "Except ye be converted and become as little children," etc., he included this whole group of Spiritual qualities, from the humility of Godliness to the docile and receptive character of Spiritual Insight and Aptitude.

The importance of this faculty of Aptitude to the teacher, as well as in the scholar, will be at once apparent. If the teacher has large Aptitude, he will readily adapt himself to the mental and temperamental conditions which, by Spiritual Insight, he may clearly discern. If these faculties in the teacher are under subjection to the Propensities, as by nature they will be, he will not be able through them to satisfy his desires in teaching, and their exercise will be continually obscured and perverted, leading him to believe in sensuous results; but if the Holy Spirit dwells in his mind so that these faculties, in harmony with Love, Meekness, and Righteousness, lead the Intellectual Faculties, he will have peculiar success and pleasure in the work of teaching.

It is often the case that Spiritual Insight is large, giving knowledge of human nature, but Aptitude small so that there is little self-adaptation to the work of the teacher; and on the other hand it is often the case that Aptitude is large, giving readiness of sympathetic action with others or Imitation of others, but Spiritual Insight is small, so that there is little sympathetic discernment of the mind of the scholar.

THE MEDITATIVE AND THE INTUITIVE CLUSTERS.—There is an obvious contrast between the general qualitative character of those faculties of this group which are situated behind those of Godliness and those in front of it. The functions of those behind Godliness are characterized by a meditative or contemplative character. By them the mind dwells upon the truth; and they exert a retentive, conserving, and self-controlling force upon the rest of the faculties throughout the brain.

Their relations to the manifestation of the Truth may be characterized as *susceptibility*; and I have designated them as the Meditative cluster of the Spiritual group.

The functions of the faculties in front of Godliness have a character of impulse, incitement, persuasibility. By them the mind acts upon, and is acted on by other minds; and through them it is involved in the higher sympathetic relations. This quality of these faculties may be characterized as *impressibility*; and I have designated them as the Intuitive clusters of the Spiritual group.

The faculty of Godliness is where both these qualities are united in the aspect of the mind towards Almighty God, impressible and vivified by his Spirit, and susceptible to and retentive of the Truth manifested by Him.

The Intuitive portion of the Spiritual group is first in order. It should lead the Meditative part, while the Meditative cluster, being the restraining part of the Spiritual nature, should support the Intuitive part, and hold it steadfast and just, in accord with the Truth. This is the condition for the true predominance of Godliness.

Washington was an instance in which the predominance of the Meditative part over the Intuitive was very great. He was not impressible; but he was very susceptible to the Truth.

Chief Justice Marshall was another instance; but the distinction between his disposition and Washington's was marked by the fact that he had less breadth in the cautious and secretive faculties and more fulness in the Intuitive faculties, though these were subordinate to the Meditative. He was more impressible than Washington.

Franklin was an instance of nearly equal development of both parts, but the central faculty of Godliness, or Veneration was less prominent in him, and followed rather than led the other Spiritual faculties.

Walter Scott was an instance of predominance in the Intuitive part over the Meditative part. And there was an extreme development of the side clusters of Intellectual faculties upward, towards the Intuitive part. He was more Impressible; but in him Veneration was well developed and consorted with the Intuitive part.

Lord Bacon was an instance of extreme predominance of the Intuitive over the Meditative which was deficient, and in him Veneration was still more subordinate to the Intuitive part.

COMPOSITE ACTION OF THE SPIRITUAL FACULTIES.—God is known only by the Presence of God in the Soul; that is to say, by the energizing power given to the soul and manifested through the Spiritual Faculties, when the human will is subjected to the Divine Will.

This plenary influence of the Holy Ghost, the Spirit of God, is not a fact of Intellectual deduction or induction, although its existence is verified by the Intellect, objectively. It is a fact of Consciousness in the Spiritual Faculties, of which Godliness is the centre and chief. The mind does not receive Him by the exercise of an Intellectual process; Philosophy, which cannot by searching find out God, must be held in subordination, and the passions, which are self-asserting, must be restrained by humility; and when the mind is in this receptive state, God manifests Himself. The Spirit of God, the author of the Soul, quickens the consciousness into life in these naturally dormant faculties of Godliness, Brotherly-Kindness, Steadfastness, Righteousness, Hopefulness, Spiritual Insight, and Aptitude; and sheds abroad, by their activity, the light of the Truth in special gifts as well as throughout the whole mind. He thus convinces the soul of sin; because, by the awakening of these faculties, the soul sees itself in the light of Truth, and becomes conscious of the sensuous and selfish character of the faculties theretofore predominant. The spirit thus brings to light the sin of the soul, making it apparent in the consciousness,

Self-abasement results from conscious self-examination with reference to the external and internal experience of the sensuous mental life of the passions, Social and Animal; hence the utter inability of the mind, unaided, to sustain itself, in the belief of a future state, by reason of the vicissitudes of its fallen nature, from the beginning of life to its close.

Self-abasement, uninfluenced by the Holy Ghost, by reason of its meek and passive conditions, can only produce a state of utter hopelessness, even when aided by all the knowledge attainable by Intellectual facts, and all that physical nature furnishes in sensuous, material, conscious evidence, together with the sentimental dependence which the moral attributes impart. Hence to possess an unfailing support in our present state of existence, and one which will bear us safely into that future which enlightened self-interest demands and our consciousness should realize, God's Holy Spirit must be possessed; and this spirit alone can be the life of the soul. The spirit when thus possessed manifests Himself phenomenally through the physical organization.

The true method of superiority among men is to be thus humbled. In heathen, as in Christian life, Veneration and Reverence for that which is above man is the source of Power. In Christianity, Almighty God is revealed to man as Personal Presence, and this is the only condition of Peace on Earth.

The Special gifts of the Holy Spirit. The Law of the Spiritual Faculties.

When the Spirit of God is thus received, and the Will of the individual becomes subjected to Him, He abides in the soul; and there bears witness with our Spirit that we are his children, by the means of the Spiritual gifts, as follows; first by Godliness, in which consists the incipient condition of receptivity; then by Brotherly-Kindness, giving, in love to our neighbors, the first test to know that we are moved by the Spirit of God; then by Steadfastness, through which God is known as an immutable God, and which holds the faculties of Godliness and Brotherly-Kindness to their important influence; then by Righteousness, which is the next contiguous faculty, through which God is revealed as a Righteous God, and which holds the whole soul in a Righteous Spirit; then by Hopefulness, which, when thus centered in the Righteousness of God, becomes the anchor of the soul; then by Spiritual Insight, which, when thus led, possesses the power of discernment both as to the inward condition of self, by God's good will, and by His Spirit to enter into and discern other spirits; and lastly, by Aptitude, which in its proper order as above stated, gives that childlike, suave relation in which we are fitted for all things.

It is only by the Special qualities residing in each of these organic conditions that the seven Spiritual Gifts of the Holy Ghost are imparted to men. The reception of the Holy Spirit is an entirety, involving the activity of the whole group of these faculties, and overshadowing the rest of the mind, but the resulting manifestations of conduct are in accordance with the organic conditions of development which characterize the individual.

The force of the Propensities and Intellect is centralized in special faculties, and quickened by the organs of special sense. These all tend to angular and special action, in the natural mind. The Spiritual faculties may be individually brought into combination and exert an incidental influence, by moral suasion; but, when illumined by the Holy Spirit, they are brought into united submission as a group, and necessity calls forth the individual predominance in their composite action, by virtue of the Holy Spirit. The law of the Spiritual life is, the General first, regulating and equalizing all the special activities in the individual. The law of the natural mind is, the Special first, and with diversity, superseding the General, and the virtue which comes from above.

The external manifestations of the gifts of the Spirit, in conduct, physiognomy, mien, bearing, language and expression, enable the observer to examine religion objectively. Morality, or the sentimental influence of the superior faculties, gives a certain geniality and graciousness of manner; Spiritual Life by these faculties awakened by the Holy Ghost gives the fruit of the Spirit, "Love, joy, peace, long-suffering, gentleness, goodness, faith, meekness, temperance."

Religion is Absolute. It cuts loose from all entanglements, and restores the soul to its own individual direct conscious responsibility to God.

PERVERSIONS OF THE SPIRITUAL FACULTIES.—The *wars and conflicts* which continually prevail among nominally Christian people, result from the predominant sway of the Propensities, which lead and subordinate the Spiritual Faculties.

Sectarianism is the natural result of the ignorance of men and their unwillingness to receive the Truth, which is one and entire, and can only be received by that humble and contrite state that is the sole and incipient condition upon which God has promised the gift of the Divine influence. Hence men without that contrition and consciousness of their inability to know the truth, contend about religion under the influence of the Social Propensities, by the aid, chiefly, of the Intellectual Faculties; and persist in their own personal pride of Self-Esteem, that being the highest of the Propensities. Every division which has marked Christianity is due more or less to these causes. If the Spiritual Faculties predominated in activity in the church, unity in love would appear, and the evangelization of the world would be rapid. Men would accept the Spiritual law of Christ, even more willingly than they now do the laws of physical science.

What is called *Animal Magnetism* consists in the suspension of the physiological activity of the Propensities, leaving the Intellect under the influence of the faculties of the Spiritual group in their natural state of blind sentiments. In this state, the mind has no activity except a reflective exercise, and becomes in a degree subject to the volitions of another, whose Propensities are in activity.

The phenomena of "*Spiritualism*," which cannot be denied, and yet cannot be explained by any of the principles taught as Mental Philosophy, and which therefore remain a mystery to the mass of intelligent Christian people, may be understood when examined in the light of the facts of Mental organization which I have endeavored to set forth. The Spiritual Faculties were designed by the Creator as the medium of His Divine influence upon the whole mind. It is through them, and the Church of Christ, which is to keep them alive, that we may receive the Spirit of God; and they have a power over the whole being, and a sympathetic power over the minds of others, which is peculiar to themselves. It is by the abnormal and erratic influence of these faculties, exerted generally by several persons combined, that the "manifestations" are produced. This exercise of the Spiritual Faculties is usually unconscious on the part of the mediums or persons engaged; for they are passive, as it were, unless awakened by the Holy Spirit from their natural state, and, thus rectified by Godliness, Steadfastness, and Righteousness, brought to predominate in activity over the Propensities and the Intellect.

"Mediums." The Restraining Faculties and the Law of Association.

Those who are known as "mediums," as a class, have the Spiritual Faculties organically predominant in size, especially in the Intuitive part, and possess a natural force in those faculties, not induced by the influence of the Holy Spirit, but merely by this natural predominance in quantity, and without due restraining influence, either from Steadfastness and Righteousness, the other faculties of this group, or from the restraining faculties of the Propensities, Cautiousness and Secretiveness.

For this reason, the practice of spiritual manifestations is demoralizing and exhausting, tending to a loss, both in the Spiritual Faculties and in the Propensities, of physical command and control, even in the natural order of social life. From the same want of restraint comes the abandon and license, to which this perversion of the noblest faculties tends, in practical life.

5. THE RESTRAINING FACULTIES.

RESTRAINT.—Life is action. It is manifested phenomenally, in movement. Thus Man is enabled to take cognizance of it, through the senses, by the Perceptions. He naturally takes interior cognizance of his own life, by the sensuous mental disposition of the Propensities, Social and Animal, as other animals do. This cognition is made Philosophic through the Conceptive and Combinative forces of the Intellect. Language or Logic is the mode of representing and communicating the processes of these organic instrumentalities, in a definite form.

Physical science teaches us that Life must be understood in its broadest sense, and it is not necessarily organic, although it is always localized, and, in the animate world, is manifested in organic forms. The forces which were once regarded as the vital principle are now seen to be the physical forces which appear in so many cor-related forms.

Restraint, is either organic or inorganic. In the inorganic form, that is, in the separation or the decay of matter without centralized design or localized life, it is the inertia by which force becomes subject to organic conditions and is made persistent.

The Physiologic activity of the organization of man becomes conscious, by the interposed influence of the faculties of Restraint,—Cautiousness and Secretiveness, Steadfastness and Righteousness.

Practical success in life depends upon how much you can manage what you have got, and this involves an organization of conscious managing, mental capacity; and this power of direction, control and prudence in worldly affairs is the function of the influence of these faculties. They are of fundamental importance for success in Education, as well as in social and business life, and in Religion.

ASSOCIATION OF THE FACULTIES.

In a previous letter I spoke of the associated activity of the faculties, and of the error of Phrenologists in regarding each faculty as having an activity by itself. The whole mind is to be regarded as one; and although by analysis we are enabled to individualize faculties, and theoretically to consider them separately, we must bear in mind that they are inseparably associated, both in activity and in development. The faculties are the individualized functions of convolutions structurally associated. The convolutions of the brain, separated by anfractuosities, indicate to us the distinguishable localities appropriated to nervous activity having certain characteristic mental qualities, and these mental qualities we term faculties; but the convolutions to which we are led to assign these faculties lie together, each one being, in some direction, connected with others contiguous to it, and all being but modifications, in form, of one continuous inward structure. Both the exterior contact of adjacent convolutions, and the interior continuity of their substance, give a certain association to their activity; so that, even when we individualize them, and speak of them as distinct faculties, they must be regarded as acting with, and re-acting on each other, in their clusters and groups.

The faculties of each group tend to ensphere themselves together in that group; and, where there is an unequal development, this tendency shows itself in the form of the head, by the extent of the region occupied by the group which has a decided predominance; either the Propensities, through their natural inherited force, increased by the development given by indulgence, or the Intellect by inherited structure, or by education, or both, or the Spiritual group. When the Propensities are predominant, the development of brain caused by indulgence of the passions is found in the downward direction, making the base of the brain large and full. When the Spiritual faculties predominate, the tendency of development, on the contrary, is upward, giving height to the head and fulness to the crown. When the Intellectual faculties predominate, the development is in the region of the forehead, giving bulk in the front part of the head.

When either of the three general classes of faculties thus individualizes itself as the predominant group, it often possesses a peripheral expansion of a cluster of the group standing out in a curve, departing from a symmetrical relation with the other two groups, so as to make the outline of the head look as if it were composed of adjacent arcs of circles eccentrically placed.

So also, if there is a great combined activity, and consequent development, in a particular cluster or congeries of special faculties in any group, as compared with other faculties contiguous to them, this cluster or com-

Mutual influence of associated faculties.

bination of organs is indicated by a similar peripheral expansion standing out in the same manner, and individualizing its character in the form of the group. And the particular faculty which leads the others in such a combination, and predominates over the others, forms the point or summit of the prominence, and gives the characteristic feature of the whole group to which it belongs. In the same way two or more faculties of separate groups often combine, producing mental manifestations of a correspondingly mixed character, either intellectual-spiritual, or intellectual-social, or social-spiritual. The present standards of moral, spiritual, and intellectual truth among men are of this mixed character; and their diversity corresponds to the combination from which they arise.

It must, however, always be remembered, that by reason of the consorting of the organs of the brain, the one which is thus projected into prominence often appears in a place somewhat different from the ordinary location, as marked on the bust, seeming as if removed, either upward, downward, forward, or backward.

Since the three groups, in their combined form, conspire together, one of them taking the lead, and shaping the head so that its own general character is predominant, the leading organ in this group will, therefore, stand predominant over all the rest of the brain, and give its own peculiar character as the chief feature in the conformation of the whole head. It may be said that the predominant group rules or characterizes the mind, the predominant cluster in that group leads the group, and the predominant faculty in that cluster leads the cluster, and through it, characterizes the group and the action of the whole mind.

Such a leading faculty, whatever it may be, is of course more or less modified, in its character, by the qualities of the faculties with which it is chiefly surrounded and associated, and especially by the character of the group to which it belongs. If it is in the Propensities, it gives to the general mental character a more vivacious action, and to the whole person more varied and pronounced physiognomic and pantomimic indications; if in the Spiritual Faculties, a more calm and meek action and expression; if in the Intellectual Faculties, a more conceptive, combinative or perceptive action or expression.

In the same manner, one of the faculties of each of the *subordinate* groups or clusters leads or predominates in its own group or cluster, and lends its peculiar phases to the character given by the leading group and faculty, thus contributing to the general make-up of the mental phenomena.

This is the general rule of ordinary and orderly development; but it is often the case that special faculties in the group of the Propensities, even when that group is not predominant in size, have so marked a development, as not only to lead their own subordinate group, but to exceed in influence the leading organs of the other groups, and give to the person a peculiar disposition and pantomimic expression.

Significance of organs whether prominent or laterally expanded.

The order in which the faculties stand among themselves within each group, and compared with the other groups, presents the conditions upon which the characteristic difference between the minds of different individuals depends, so far as these differences in the individual character arise from the organic shape of the brain.

All these conditions appear in the general form of the head; the most predominant group, cluster, and faculty, in their peculiar composite form, giving the most salient features to the outline of the head; and their relative prominence, and the peculiar conformation of the subordinate parts of the brain, indicating how far they characterize each other; so that the whole form is thus significant of the general mental organization.

By the posture of the organ in such conjunction, tending upward, downward, forward or rearward, the special quality or tone of the action is manifested, whether more sensuous or more *spirituelle*, more intelligent, or more passional; and this generic modification depends for its particular quality upon the function of the organ contiguous in this conjunction. And in the same way the conjunction and posture of a cluster or group, with reference to the contiguous clusters or the other groups, gives to the observer the leading phase of the cluster or group.

The expansion of a predominant organ modifies the shape of its neighbors. Where faculties are habitually associated in their activity, the organs combine more closely in their position. A predominant organ not only presses upon the weaker organs, but draws to itself and sometimes overlays the contiguous organs which are most active in connection with itself, presenting to the practical observer of the external shape, the appearance of a merging of the two convolutions.

From this it results, that the mere special prominence of that part of the surface of the head which is assigned upon the bust to any particular organ, does not necessarily indicate the predominance of the corresponding faculty, individually and by itself. But the necessary and proper bulk or form which the special faculty assumes must be taken into account, and the effect of the predominance of the faculty, in modifying the position, shape, and activity of the adjacent organs, must be considered, in estimating the significance of the external shape of the head. For instance, if in any head, the region marked on the bust for the location of the organ of Godliness appears to be somewhat sunken in the centre, in contrast with the contiguous organs around Godliness, the cause may be, that this organ is deficient, or it may be there is a large expansion in the region of Brotherly-Kindness, which lies in front of it; or if Brotherly-Kindness, lying in front, and Steadfastness, which is behind it, are both large, it may be that they may have drawn the organ of Veneration or Godliness upon their respective sides, so that it may be low in the middle, but broad, and upon each side well developed towards the more predominant neighboring organ. Thus, in the profile view of George Washington, the retreating upper forehead does not in fact indicate that he was lacking in Brotherly-Kindness; but shows that the organ of Godliness was so large and predominant, that the organ of Brotherly-Kindness was drawn to it and consorted with it. Again, Self-Esteem was in him the largest organ among the active forces in the Social Propensities, and this organ, consorting as it did with Steadfastness, made him such an excellent commander, and gave to his head that peculiar peripheral expansion in that region and upon the line between Steadfastness and Godliness.

To illustrate this in a more general way, if you look again at the same profile view of the bust, you will observe that from the upper verge of Brotherly-Kindness over to the upper verge of Self-Esteem is nearly the arc of a half circle, having for its centre the ear; but this arc rises above the rest of the circumference of the head. The phrenological explanation of this is, that Steadfastness and Godliness were the chief characteristics of his mental life, and that they drew to themselves, respectively, the organs of Self-Esteem and Brotherly-Kindness, and the latter qualities consorted with and entered into Steadfastness and Godliness. The force of the outward circumstances and influences centralized upon him as the superintending chief of the Revolution, and a predominance of the Meditative cluster in his Spiritual group of faculties, and a disposition to yield to no external influence, save that of patriotism alone, considered in connection with his passive temperamental qualities, constituted him the special ordained character, which God in his Providence selected for the results which were to be attained.

The Spiritual Faculties (being by nature in themselves passive) found their support and the force of their activity in the faculty of Self-Esteem; and, as the vital forces reside in the Propensities, and Self-Esteem is one of the highest, and in him was the strongest of these faculties, it produced, in combination with Steadfastness, the dignity and gravity of character which he manifested in his military and official life, while he predominance of Godliness or Reverence, made him equal and accessible to all.

This characterizes truly the mind of Washington in these respects.

The predominance of the upper part of the head in this profile view, would have been termed by Dr. Gall a bump. It is always to be remembered that such a prominence or bump, is not the indication, merely, of strength in one faculty, but rather of an association of faculties, consorting and acting together.

Other views of the same head which are presented, show the lateral organs which were also drawn to these central faculties of Steadfastness and Godliness, and consorted with them.

The relative predominance of the groups, is always the first element of form to be considered, or we may be misled in estimating the comparative development of faculties in one group by disregarding the great influence of another group.

7. LOCATION OF SPECIFIC ORGANS

As I pointed out in the preceding letter upon the temperaments, the first step in Phrenologic observation of character is to estimate the quality of the temperament of the person, and form a distinct idea of the relative order in which the four leading systems stand, and their relative strength in the combination.

The second step in Phrenologic observation is to see which group of faculties predominates. At this point we have come to conditions which involve geometric form; and although absolute measurements are not practicable, estimates of form must be made with an endeavor to attain the same kind of definiteness and accuracy, that actual comparison with a uniform standard would give. To form just estimates of the contour of the head, and the comparative bulk of its parts requires some experience in observation; but, although some persons possess a special aptness for it, any person of good intelligence in regard to form and size, by attention and careful thought, may attain a sufficient degree of accuracy in judgments of the general cast of character. The best aid for the beginner will be to study the form of the head in comparison with that of a sphere, or, taking the same idea in a simpler form, to study the outline in comparison with the arc of a circle.

Phrenologists have sought to define the location of faculties by superficial measurement, and have endeavored to reduce the results of observation to mathematical formulas, stating the comparative quantity of the influence of each faculty according to a numerical notation.

But, from what I have already said, you will understand that such methods of superficial measurement, however much they might assist observation in respect to details, would rather mislead the beginner by turning his first attention to seeking for superficial prominences or depressions, which are necessarily somewhat uncertain in their significance, and are, at best, of secondary importance. It is the *general form of the head*, which must first be clearly individualized.

Estimates of character, formed without distinct reference to bulk and the characteristic direction of development, and depending primarily on the observation of local prominence, or depression in the surface, whether measured and localized by the eye, or by feeling the head, or by the measuring tape, give some knowledge of character; but correctness in estimates thus formed is only attained by long experience, which enables the observer to be guided by the impression made in his own consciousness by the subject, and to be very guarded in interpreting the special salient features of the head.

General form of the cerebral instrument.

The obvious fact that the process of making observations in this way is necessarily peculiar to the manipulator, and not accurately definable to others, has caused the practical application of Phrenology to be regarded by many as empirical, and like fortune-telling.

What is needed, is a delineation of the whole configuration with reference or in comparison to an unchanging geometric standard, which will be both accurate and indisputable.

For the ordinary purposes of the teacher, however, all that is needed is a training of the Perceptive Faculties, or in popular language, a well trained eye, such as will enable him to recognize with clearness the characteristic form which the head of any child possesses, in the same way as a ship builder recognizes and compares the characteristic lines of vessels, or an engineer measures the lay of the land with the eye, before he places his instruments.

In other words, the head is to be defined, not by superficial measurements of distance, but by a comparison of the curves and diameters which will indicate the bulk, and the direction of development, whether upward, forward, lateral, backward, or downward.

Each part of the brain has thus its character, depending on the shape, the degree of fibrosity and the length of fibre.

When we have ascertained the general features, as I have before explained, and the direction of the greatest development, we have indications by which, taking all other things into account, in the case of adult subjects we may judge in what direction the longest fibres run, and in what part is the greatest fibrosity; which results from the predominant activity of the mind. We must carefully distinguish whether it is sensuous or mental, Spiritual and illuminated, or Animal and Intellectual. These are the general phases to distinguish the qualitative activity of either of the groups of mental manifestations.

PRACTICAL METHOD OF DEMONSTRATING THE FACULTIES.—The first point to be observed in such a measurement is the position of the opening of the ear.

Since the head is eccentric or irregular and variable in its form in every direction, its centre cannot be determined by relation to the surface, or bulk. It is the very object of our measurement to ascertain the special variation of the circumference, by measuring from a fixed centre. It is found by observation, that in an orderly developed head, the opening of the ears is midway between the front and the back; and as the two sides of the head correspond to each other in faculties, the middle of a line drawn through the head from the opening of one ear to that of the other will be the point from which, as from a centre, to measure the relative development in any direction. The base of the semi-circles in the dia-

Indications resulting from various cerebral forms.

gram, it will be seen, is a straight line which intersects this axis. This base line is horizontal, both in the front view, given in the second diagram, and in the profile view. It must however, be observed, that in the natural position of the head, the base of the brain is not generally horizontal on a plane in the direction shown in the profile view, but inclines downward toward the back of the head; the lower Intellectual Faculties, in front, usually ranging higher than the lower Propensities behind, being two planes, as represented on the diagram of Washington, higher than the Propensities.

By keeping the eye on the half circle in the outline, the relative prominence of the organs which lie along the outline of the head in that view is readily apparent. The radius of the sphere, in the full size of it, is seven inches, the diameter being fourteen inches; and a scale on the base of the half circle, in the diagram, upon which the inches and portions of inches are laid off in the proper reduced proportion, affords an exact measurement by which the development of any organ in the circumference can be ascertained, the centre of measurement being always in the axis passing through the opening of the ears. No head will be found fourteen inches in diameter; but some heads approximate to seven inches from the axis forward to the front of the organ of Individuality, the ears appearing in such case to be very far back; and on the other hand, some heads approximate to seven inches from the axis back to the outline of the organ of Philoprogenitiveness, the ears in such case appearing very far forward. It will be observed that the head of George Washington, as represented in either diagram, approximates to the upper part of the circle within about an inch and a half, as marked on the scale.

Hence it may be said, in the ordinary observation of Phrenologic form in the profile view, if the ears appear very far forward upon the sides of the head, it is because the development of those Social Propensities which lie behind, in the central part, is much larger than than that of the Intellectual group which is in front; if the ears appear very far backward, it is because the central portion of the Perceptive cluster of the Intellect is prominent; or if the greatest extension is upward, the greatest height being directly over the middle of the axis, it shows that the central portion of the Spiritual Faculties where the two hemispheres lie contiguous to each other, predominates.

If the head should be placed within a sphere, so that the centre of measurement should coincide with the centre of the sphere, the parallelism or the divergence of the surfaces of the head and the sphere would afford a means of estimating the development of the brain in every direction.

It must not be supposed, however, that this hemispherical form stands as the delineation of a true order of mental development. It is only a rule of measurement, with definite distances from a given centre, so as to define in all shapes, in what region the quantity of brain lies. But a

Indications resulting from various cerebral forms continued. Influence of Temperaments.

head which should correspond to the curve of the sphere, would be well nigh the worst mental disposition. It would present the great strength of the mind in the appetites and passions, and the lower mental faculties. The person whose head corresponds to the contour of the circle is of an earthly nature. The higher order of development is that in which the head, like that of George Washington, predominates in the Spiritual part, rising high and round in the upper part of the area.

If the surface of the head, compared with such a sphere, was found to be parallel with that of the sphere, it would indicate that the brain formed, approximately, a hemisphere, in which the centre of bulk coincided with the centre of phrenologic measurement; and from this conformation we should deduce a certain corresponding mental character. It might be that another head would be equally spherical in its form, but that the development giving this form, might be rather in front of, or behind, the centre of measurement. In this case, the centre of bulk of the brain would be thrown forward of, or behind, the centre of the ideal sphere, and the surface of the head, though spherical, would not be parallel with that of the sphere, but approach it more closely in one direction and recede from it in the other. From the conformation of the brain thus indicated, we should infer a different mental combination, although the general shape of the head would be the same.

If again, as in well-formed heads, the development upwards exceeded that either forward or backward, making the outline of the head more or less oval than circular, the degree of this upward development would be indicated by the contrast between the curve of the head and that of the sphere; and, as before, whether this upward development was vertical or forward, or backward, would appear, by observing whether the perpendicular axis passing through the centre of the bulk of the head coincided with the perpendicular axis of the sphere, or was thrown forward or backward of it.

Another modifying influence to be considered, in estimating the variable locations of the convolutions, is that of the temperaments. The bodily functions, in which the nervous forces are often engrossed, have respectively their reciprocal relations in specific faculties of the mind; and hence, in general, it results that the size, location, and activity of certain faculties are to some extent modified by the existing temperament. This may be discerned especially in the characteristic mental and temperamental difference of the sexes.

A third disturbing element is the variable thickness of the cranium, and the existence, frequently, of a cavity between the external and internal layer of bone, over the nose and behind the eye-brows. The difference in thickness of the cranium in different parts is but slight, and does not sensibly affect the general form of the head. The frontal sinus is sometimes so large that it might materially mislead the observer's judgment of the development of the Perceptive faculties. It is said that this sinus does not exist in children under the age of twelve; and it will not as a general rule be found except in connection with the Bilious-lymphatic temperament.

These minor disturbances are to be considered in Phrenologic observations; but their existence does not impugn the basis or the just deductions of Phrenology.

A development of the Propensities downward making the greatest breadth in the base of the brain, which in such case projects outward, and produces a ridge more or less apparent, indicates a concentration of the forces of the mind in the lower appetites and passions, with sensuous, vivid, and acute activity, and shows that physical conditions take precedence. A development of the Propensities in the direction of the Spiritual Group, the head enlarging upward and receding at the base, being more contracted there, and sometimes presenting a furrow or sunken band upon the head at the base of the brain, indicates that the force of the Propensities, instead of being acute and physical, are rather passive, sentimental, and by conspiring in their activity with the upper range of faculties, have a moral, Intuitive, or Meditative tone.

This is the first phenomenal aspect that is to be observed in determining the general phrenological features of any character. Before proceeding to examine the development of special characteristic faculties, the outline of the head is to be thus looked at from the organ of Individuality lying in front, around to that of Philoprogenitiveness behind, so as to present to view successively each faculty lying at the base of the brain. In adult subjects the cerebellum, which lies in part below the base of the circle, must be particularly observed. Size in this part of the encephalon or activity of its functions, indicates a precedence of the sensuous nature. Before the general character has been thus ascertained, the observer is liable to err in judging of special faculties, for, a correct knowledge both of their location, and of their quality as affected by associated action, depends upon the knowledge of the centre and direction of development of the whole brain.

The salient phrenologic features of the head are, of course, most plainly seen when the head is viewed in such positions that the salient points are brought into the outline; but if the attention of the learner were confined to the outlines, there would be a tendency to study first, these special points, without sufficiently regarding the fundamental facts of size, length, breadth, height, and the direction of the characteristic development estimated from the phrenologic centre.

In order to direct attention more strongly to these primary aspects, I present a number of diagrams in which the head of Washington appears in successive positions, surrounded in each case, by the same circles, great and small, lying in the circumference of a sphere, the centre of which coincides with the centre of measurement.

The vertical lines are the arcs formed by great circles which intersect each other upon the perpendicular axis of the sphere.

Description of diagrams. Mental characteristics of Washington.

The plane *a* A *o*, which passes through the brain from front to back, divides it into its two hemispheres. The plane *h* A, which passes through the horizontal axis of the head from ear to ear, divides each hemisphere into a front and back part. Each of the four portions, into which it is thus divided by triangular surfaces, approaches the form of a spherical wedge or ungula. The position of the ungula which contains the front part of the hemisphere is indicated by the half lune *a* A *h*; and that which contains the back part of the hemisphere is indicated by the half lune *h* A *o*. The first question to be asked in reference to the form of the head is, which of these portions contains the greater bulk of brain. To form an opinion on this question, it is necessary to determine, first, the centre of measurement, and to consider then, not only the length of the head which is shown in the profile view, but its breadth throughout, which is only seen in a front or back view.

Groups.—It is obvious, in viewing the head of Washington in the frontal aspect that the most salient feature is the great width, that is to say, the large quantity of brain in the breadth of the head. The extreme breadth is in the region of the faculties of Cautiousness, making these with Destructiveness and Secretiveness, which consort with them, the significant features of this view. This cluster of faculties is delineated as having an upward development, as will be seen by the less breadth which marks the base of the brain in that region. The influence which caused this upward development of these faculties must be looked for in the profile view, in which, as we have before said, the salient feature is the great quantity of brain in the region of Self-Esteem and Steadfastness.* This combined feature we should describe by designating Self-Esteem and Steadfastness as the predominant faculties, they drawing Cautiousness backward and and upward, giving to it, and to the lower faculties of Secretiveness and Destructiveness, which consorted with it, their own superior characteristics, and presenting in the external form a contour significant of this peculiar order of combination. In this way, in every head marked by any special development, the region which contains the largest quantity of brain is first to be sought for, and it then will be found that the form of the faculties surrounding it in the same group, and often throughout the whole head, are moulded as if they were drawn to, and subordinated by these convolutions of largest relative size.

By this delineation of the faculties named in the group of the Propensities, and the form under which, in his head, they are blended with the faculties named in the Spiritual group, and contradistinguishing the natural force of the Propensities from the mere qualifying or modulating character of the moral sentiments, we see defined the central quality of his character; a self-determined and inflexible, cautious, secretive, and executive will, vigorously defensive, and destructive to those who opposed themselves to the cause of his country's freedom. And the quality of

Boundaries of Groups not fixed. Activities of the Propensities must be regarded.

mode of the manifestation of this mental character, we see defined in his temperament. The peculiar difficulties which beset him, arising both from friends and from foes, presented obstacles which could only be surmounted by great strength in the passive elements of force, and by the lapse of time. Washington was not driven by an over active brain, or nervous temperament, nor a stiff or metallic will of his own, nor a hasty and inconsiderate, spasmodic, Sanguine temperament; but his great force of mind placidly awaited the exigencies of affairs, and was moved by paramount necessity to calm but irresistible action. Nothing but success, could suspend or deflect the energies thus aroused.

In making observations of the relative predominance in size and activity of either of the three groups, it must be borne in mind that the boundaries between them are not arbitrary and fixed limits. They are indicated upon the diagram of the head of George Washington at the points where, in my judgment, they existed in him; but the position varies in different heads, according to whether the greatest quantity of brain is in one group or the other; and in the examination of any head, the location of the boundary must be first determined.

Thus if the Propensities greatly predominate, especially in the upper part, they may trespass upon the usual region of the Spiritual Faculties, carrying the boundary between these two groups up to a higher point. So if the great predominance is in the Intellectual Group, the faculties of that group will trespass upon the forward boundary of the Spiritual group and press it backward.

But it is not enough to observe predominance in *size* alone. The activities of the Propensities are to be regarded, because the vigor and animal force reside there. This force is often enervated by satiety and a profuse supply of all physical wants; while a degree of necessity awakens it, and leads to activity and energy in the Intellectual Faculties. But the Propensities tend, when under great necessity, to overlay the other faculties and assert themselves, especially those Propensities which lie contiguous to faculties in the other two groups, and they will do so, unless the moral force of the Spiritual Faculties, or a diversion of the activity into the Intellect or into the body by physical exercise, supersedes them. The indications afforded by the tones of voice, the mien and manner, habit and posture of body, the expressions of the countenance, and the temperamental conditions, all are important, in estimating the relative predominance of the group.

Thus the man whose mind has been changed to a Spiritual nature will manifest, externally, that mild and meek character which is a general index of the predominance of the activity of the Spiritual group. So the man whose Intellect has received the education of a proper collegiate course, and has experienced the discipline of society, will manifest in expression and manner, the results of that predominance of the activity of the Intellectual group, which education only gives.

Indications resulting from predominance of various Groups and Organs.

Thus also the self-educated man, though he may be unpolished in manner and in classical expression, the result of defective social culture, will yet manifest similar evidences of Intellectual activity, though in a less attractive form. Self-education, when conducted without the refining influences of polite society, results in a rugged, unmethodical development of the mind, manifesting itself in a corresponding exterior; whilst a thorough classical education, such as our Seminaries of learning impart, combined with suitable social advantages, affords a systematic mental discipline which is apparent in the characteristic expressions and polished deportment of leading literary men, and this is more apparent in Europeans than Americans.

The ordinary position of the head is an indication of the predominance of activity. If the Intellectual group predominates, it will be observed that the head hangs forward. If Self-Esteem predominates, the head will not only incline forward, but the chin will be drawn in toward the throat. If Godliness predominates in a moderate degree, the tendency is to an upright perpendicular position of the head. If the Meditative part of the Spiritual Faculties in the back part of the upper region of the head predominates, the tendency is to incline the head forward, so as to give those organs apparently a higher position than they would otherwise have, on account of the will being centralized in this cluster; and if the Intuitive, in the upper and forward part of the head predominates, the tendency is to raise the countenance to look upward, and thus to give these organs an apparently higher place. The side organs of the Propensities give a different pantomimic expression. Thus, if Destructiveness is predominant, the chin is thrown forward, and the lines of the mouth are made more marked, and the lips firmly set. If Secretiveness and Cautiousness are large, the characteristic pantomimic expression is a side-long movement and position of the head.

These indications, however, are less observable in children than in adults, because of their undeveloped, sensuous and volatile character, and therefore I do not here describe them in detail.

CLUSTERS AND FACULTIES.—Along the outline which is presented by the profile view of the head, lie those organs which run through the centre of the head, where the right and left hemispheres of the brain are contiguous to each other.

These are designated on the diagram:—they are Individuality, Eventuality, Comparison, Brotherly-Kindness, Godliness, Steadfastness, Self-Esteem, Inhabitiveness, and Philoprogenitiveness.

Let us first observe the regions of the Perceptive cluster and the Conceptive cluster, respectively. In the lower part of the forehead the frontal sinus may give the appearance of great acuteness of special organs. The observer who has first regarded the development of the group, will be prepared to make the necessary allowance for this interference with external form.

The two faculties or organs of the Perceptive cluster which appear in the circumference are Individuality and Eventuality, and their relative prominence is at once seen. The location of the organ of Form is also marked upon the diagram, but the development of this organ is not indicated by prominence, so much as by width between the eyes, and it is best seen in the front view of the face. The one faculty of the Conceptive cluster which appears in the circumference is Comparison. By thus individualizing these organs, and comparing their outlines with each other in the circle, we see in what region lies the preponderance, and what faculty predominates in this view of the head, and we discover, in the general shape, how these faculties consort with each other.

Passing onward from the Intellectual to the Spiritual group, it will be observed, that, in this diagram, the faculties of the Spiritual group have a peculiar form, the Intuitive part retreating very rapidly from the organ of Comparison, as if the organ of Brotherly-Kindness had left its proper place. This gives a very retreating forehead. The Intuitive cluster is quite deficient, the head however being high in the region of the Meditative cluster. This indicates the strongest characteristic of George Washington, and some of his habitual sayings corroborate the observation. "I meditate to pass the remainder of life in a state of undisturbed repose."

Passing the line of demarcation between the Spiritual group and the Social and Animal Propensities, we see the largest quantity of brain to be in the region of the organ of Self-Esteem, this being the largest faculty of the Social and Animal group in his character. From this, the organs diminish in a proper order down to the base. Home and children had less influence upon his character than his dignified position.

Let us now turn to the front view of the head, which is presented in another diagram.

Here, the front part of both hemispheres of the brain appear, giving the general width of the head. This view shows all of the Intellectual Faculties, the Intuitive part of the Spiritual group, and a part of the Animal group. Upon one of the hemispheres, are delineated the names of the clusters, and in the margin, are given the names of the groups in which these special faculties are associated. The names pertaining to the Intellectual group are designated by the use of small letters. The names of the organs of the Combinative Faculties on the bust are obscure in the facial view, but these names and the locality of the organs can be distinctly seen upon the profile view.

By looking a moment at the position of the organs delineated upon the profile view, it will be seen that the plane of the circle, $h\,A$, as placed in the facial view, intersects the head in the front part of the organ of Destructiveness, passing between the organ of Acquisitiveness in the Intellectual group, and that of Cautiousness in the Propensities, and through the middle of the organ of Hopefulness, up to the centre of that of Godli-

Front view continued. Character of Washington as indicated by cerebral form.

ness, and the line of intersection passes down, in the same way, through the twin faculties on the other side of the head. Alimentiveness is indicated upon the diagram, because it is alone, below Destructiveness, and the foremost of the Propensities, though it is not exactly in the plane of the circle.

By this view the observer will see where the greatest breadth is developed—whether it is in this central region, as in Washington, or forward of it, in the Combinative cluster of the Intellectual Faculties, and the Intuitive cluster of the Spiritual Group, giving the fullest outline in the region of the line $g\ A$, or backward from the central region, in the restraining faculties giving the fullest outline in the region of the line $i\ A$; and again, whether it is upwardly developed as in Washington, giving the greatest breadth in the higher Propensities and the Spiritual group, or downwardly developed giving the greatest breadth, (whether forward or backward,) in the lower Propensities, in the base of the brain.

To complete the view of the individual faculties in detail, the head should be looked at in every successive position necessary to bring each organ, from Individuality in front to Philoprogenitiveness behind, into the outline.

The diagrams, of which engravings are given, are selections from a complete series of 15 Photographic views of the Bust, one being taken from the plane of each of the great circles, from a to n.

For practical purposes, however, views from two or three positions will disclose the salient features of the character.

The salient features of the character of George Washington, indicated by the cerebral form, in connection with the temperament, may be thus stated. The commanding mental qualities indicated by the profile view No. (1), which presents Self-Esteem, Steadfastness and Godliness as the leading and predominating characteristics of his mind, in connection with the breadth of the head, upwardly developed, in the region of Cautiousness, Secretiveness and Destructiveness, and in combination with the Lymphatic, Sanguine, Nervous conditions of the temperament, gave him both that mental and temperamental state by which he was distinguished during the Revolution. The adverse and apparently inextricable difficulties surrounding the American Colonies aroused the dormant energies of this youth, which nothing but overwhelming necessity could awaken.

The mind of Washington being strong in the central Perceptive Faculties of Individuality, Locality and Eventuality, gave him capacity for that special knowledge of present facts necessary for him, as Commander-in-Chief, and in combination with his predominant mental force in the faculties of the Meditative cluster, giving him a silent, prevenient and personal wisdom, illuminating the Intellectual Faculties thus led by the Perceptive cluster. With this mental organization, the negative qualities of his temperamental constitution gave the great reserved power, which constituted the central element of his genius.

The Intellectual discrimination which depends upon observation of the person, consists in the observer's regarding minutely the objective facts, of form, size, color, tone, etc. This is done by the Perceptive faculties, and, in connection therewith, by faculties in the Combinative cluster defining and individualizing the composite result arising out of the combination of these objective facts as seen in the temperament and the form of the head. Success in deducing the character from these data depends upon the gift or faculty of Insight.

My duty in Phrenologically describing character, thus observed and discerned, involves the exercise of the faculties in each of these three clusters, the Perceptive, the Combinative, and the Intuitive.

SIZE AND ACTIVITY.

SIZE NOT AN ABSOLUTE CRITERION.—In estimating the character from the form of the head, two fundamental principles must be always kept in mind:—The first is, that structural predominance, or the mere relative size, indicates capacity, but not, necessarily, actual manifested qualities. The second, which is connected with it, is, that Temperamental conditions modify the mental manifestations.

The general law stated by Phrenologists is, that other things being equal, size is the measure of power. This is a fundamental principle, and too much importance cannot be attached to it. But it is necessary to know what are the "other things," equality in which is the condition of applying this axiom, and how we are to ascertain their equality, or what allowance is to be made for inequalities. Those who have treated the subject controversially have sought to justify a rejection of the general law, because the investigators who asserted it had not elucidated the qualifying circumstances which must be considered.

The most important of these conditions may be stated as follows:

1. *Temperamental indications.* From the indeterminate composition of the organized substances, it results that two brains of the same size will not generally contain exactly equal amounts of the organic constituent elements upon which the exercise of the mental disposition depends. The brain which is the seat of mental life, is sustained in its activity by the other three temperamental conditions; hence the means of harmonizing apparent incongruities is to examine the other temperaments. By a comparison of the size and activity of the temperamental systems, we may learn the characteristic effect in modifying mental manifestations.

2. *Harmony of Cerebral Parts.* Since different parts of the brain have different functions which do not have the same relation to mental power, it results that two brains of the same size, and influenced by similar temperamental conditions, will not manifest the same degree of power, unless they are characterized by a similar proportion of parts.

Degree of Fibrous development. Activity of the Faculties is the measure of influence.

A marked difference in the proportion between the groups, in brains of equal size, will give rise to a marked difference in what is called mental power or force.

3. *Degree or Period of Development.* The degree in which fibrosity has been developed in the brain, or in any particular part of it, is another contingency which qualifies the rule, that size is the measure of power. An adult brain possesses generally more mental power than a child's brain of the same size. This degree of fibrous development, we can only ascertain approximately, by observation of external expression in the manifestations which show the habitudes of the mind.

4. *Activity as the measure of influence.*—While, under such qualifications as the foregoing, *Size* is the measure of *power or capacity; activity* is the measure of *influence*.

There are many men in whom the organs of the Intellect are predominant in size, but who are not called intellectual men. So, too, there are many men in whom the organs of the Combinative cluster, or the Conceptive cluster, are predominant in size, but whose minds are not to be characterized as Combinative or Conceptive, because the predominant activity is not in those clusters. To give a predominant Intellectual character to the mental manifestations, there must be not only the natural or structural capacity for Intellectual power, but there must be a sufficient force in the Propensities or in the Spiritual Faculties, to call the Intellect into activity, and sufficient influence of the restraining faculties to give control and direction; for, it is directed activity which is the condition of the influence of the Intellectual faculties in characterizing the general cast of mind.

Structural predominance of the Intellect gives, however, a predisposition to predominance in activity, and both these conditions of character, therefore, usually concur in respect to the Intellect. But if the restraining faculties are not consciously applied to the Intellect, and if the organization of the Perceptive, Conceptive or Combinative clusters has not an orderly development, more or less conforming to that marked on the bust, there will be less power in the manifestations of the Intellect, on account of its angular development: and the degree of this diminution will appear in the Phrenological form of the head.

It devolves upon Education to secure this power, by promoting the orderly development of the whole.

Where there is a due degree of restraint, the mind is able to utilize the best capacities, and in such minds the tendency will be to use most actively the largest faculties in the Intellect; but if restraint is deficient, the man, through lack of control, will sooner or later come to depend on others to be directed, and, by reason of the popular ignorance of the mind, he may possess large Conceptive powers and yet be employed, perhaps for life, in a vocation which exercises the Perceptive cluster, almost exclusively, and leaves the larger faculties of his Intellect comparatively inefficient. So that, though structurally, his is a Conceptive mind, practically, it will be a Perceptive mind, and *vice versa*.

Where there is an undue predominance of these restraining faculties, the Intellect cannot remedy the difficulty. The full consciousness of the Spiritual faculties must be awakened, to give the indwelling power of grace; for where this sentient knowledge is awakened, there results, first, the consciousness of what is the existing error, and, next, the disenthralment of the mind, and the subordination of these selfish powers of control to the divine will.

Again, there are many men in whom the organs of the Spiritual faculties are predominant in size, who are not spiritually-minded men. Predominance in size in the Spiritual faculties constitutes an organization fitted for manifesting the power of the Holy Spirit, if received; and a man thus constituted will be predisposed to the passive morality and natural sympathy, which the moral sentiments possess, in accordance with the manifestations of these faculties when blind, as characterized by Dr. Spurzheim; but these faculties, even when thus predominant in size, are still prone to be obscured and superseded by the Propensities which take precedence by bodily force, or by the Intellect which is awakened by the senses. They must be awakened by the Holy Spirit, to give predominance *in activity*, as declared in the Holy Scriptures.

5. *Modifying effects of Temperaments.*—The mind resides in the brain, through which it is manifested, and the quality of its manifestations is primarily conditioned upon the organic structure of the brain, although, as we have seen by a considerate application of scientific principles, it is qualified by a higher power to do what by its natural or sensuous power it can not do; but the brain is so dependent upon the bodily conditions for its forces and support, that any marked extreme in the temperamental constitution produces a marked modification of the mental manifestations.

For instance, if the faculties of Combativeness and Destructiveness are largely predominant, and the restraining faculties of Cautiousness and Secretiveness, and of Steadfastness, (or "Firmness") and Righteousness, (or "Conscientiousness") are very deficient, and if, with this peculiar structure, the person has the Sanguine-nervous temperament, the mental manifestations will be volatile, quickly affected by every thing surrounding, and re-acting immediately under every influence; and the person is more the creature of surrounding circumstances than of any responsible motives of his own. The carelessness arising from the lack of restraint, shows itself in irritation, precipitancy, and levity, unless the Holy Spirit has awakened the soul to its higher consciousness in the Spiritual faculties.

The same mental organization, with the Bilious-lymphatic temperament, would give torpidity, retirement, reticence, indisposition to interfere with affairs, and indifference to surrounding circumstances; and the person would almost rather starve than work. The carelessness resulting from deficient restraint will show itself, in this case, not in rashness and levity of conduct, but in improvidence and in inactivity.

If, however, the restraining faculties are large, predominating over Destructiveness, and the temperament is the Sanguine-nervous, the mind keeps a prudential care over itself, modifying every motive arising from the selfish Propensities, and uses the acquaintance which it has with surrounding circumstances for guidance; and this makes a character successful in very active engagements in the practical affairs of life, whether in Religion, Philosophy, Literature, Art, Commerce, Mechanical pursuits, or Politics; but with the same mental organization, with the Bilious-lymphatic temperament, the person would show a disposition to retirement, and to avoid apparent and active labor, unless under the pressure of great necessities; and would, in his retired way, do more to hoard, and amass wealth than those of the Nervous and Sanguine temperaments.

The pantomimic expression and the physiognomic form, which indicate habitual activity, are therefore very important, in connection with the shape of the head, as expressive of the mental character. With every form of activity and of temperament, is a corresponding physiognomic and pantomimic expression; showing itself in gait, tones of voice, gesticulation, and the movements of all the senses. No two persons are any more alike in those respects, than in the shape of the head and the conjoined sensuous conditions and temperamental characteristics.

THE LAW OF DEVELOPMENT.

The healthful activity of any faculty necessarily tends to its development, especially in the period of childhood and youth.

The development of the faculty may consist in an increase in the size of the corresponding organs, or in a refinement and improvement of the quality of those organs, or both.

Whether the development will be in respect to the increase of the size of the organs, or by giving it superior quality, will depend much upon the special predominating character of the temperaments and their sensuous dependence. The tendency to development in size is due, predominantly, to the Lymphatic temperament; and, as this temperament generally leads in childhood, the process of development in size is then most actively marked. When the Nervous temperament predominates, refinement rather than enlargement is the characteristic result of activity.

Modifications of activity, arising from changes in the temperamental conditions, are to be distinguished from the development of the faculty. Physiological means, applied through the temperaments, may increase the power of manifestation, and, if properly understood, will directly lead to development. Teachers often err in supposing that there is a deficiency of mental development, to be treated by exercise of the faculties, when all that is needed is the proper physiological conditions for influencing the temperaments.

The bodily conditions modify the mental action. The nervous activity is often largely diverted from the brain and special senses to the great organs of the temperaments, and the general sensuous conditions of other bodily functions, so that it may sometimes be said that the man lives in one of them, for instance, in the stomach. When the Propensities are predominant, their activities, if not drawn forth into the Intellect, are peculiarly prone to be engrossed thus in the body. If, on the other hand, the Intellect is developed in undue proportion to the body, the life is too purely mental, and there is not enough force in the Propensities.

If the system is properly balanced in this respect, the vigorous and healthful activity of the lungs and stomach receive their due share of nervous power, and the liver, under proper conditions of seasonable bodily repose, performs fully its functions, and the brain is thus both sustained by a well-ordered bodily system, and relieved, by the occasional diversions of the nervous force, from a too exclusive activity within the head. All parts of the system thus participating in due proportion in the vital action, life and symmetry is given to the whole man, so far as physiological conditions can accomplish this result. Upon these conditions, the Spiritual power must supervene, at the proper age, to give the complete conditions of perfect life and constant power.

In the earliest period of infancy the brain is of a Lymph quality. The first appearance of mental life is in the child's crying, which arises from the sensitiveness to the change of atmospheric relations it experiences. With these premises, it will be seen how peculiarly the brain becomes acquainted with the outward world through this sensuous nature. Its first disposition arises from the lungs, next from the stomach, and then from the liver by secretions. It may be said that the child is not mental until it begins to seek for its own food by the organs of smell or taste, and successively it comes to recognize its own mother, or to call her name, or to distinguish the different sounds of the words father and mother, or to know one thing from another by the sensitiveness of touch. This gradual development of mental life in the infant is a study full of instruction to one who is interested in observing its earliest and progressive manifestations. In the absence of any definite knowledge upon the subject, it may be, that the first impression of the infant originates in some faculties residing where the two hemispheres come together, which may be those of heat and cold, but it is only by Anatomy and Physiologic manifestation, that the location and function of these faculties can be established.

General law of the development of the Faculties in children. Diversities of this Law.

To a certain extent, the faculties have a natural relative order of development in point of time. Thus, mental life begins in the Social and Animal Propensities, (excepting the cerebellum.) The Intellectual Faculties are next called into exercise, and the cerebellum and the Spiritual Faculties are not predisposed to be awakened until much later. This is the general law. But when we seek for the relative succession of development among the particular faculties, we find great diversity in different individuals.

It will be observed that the forehead of young children is usually more prominent in the upper part, and more contracted in the lower part, in the region of the eyebrows. This is the result of the fact, that the forces of the Propensities appealing to the Intellect awaken usually the Conceptive faculties first. The Perceptive cluster is developed by the senses, and the quickening of their activity follows after the inquiring spirit of the mind which has been awakened by the Conceptive faculties. But in some children, the Perceptive Faculties are more early developed than the Combinative or Conceptive; some again, for instance, will recollect the mother and father by name, before they begin to walk; others will begin to walk before recollecting names. The physiological explanation of this difference is, that in the one, the special faculties of Language, (which are in the Intellectual group, and are among the lowest mental functions like that of Alimentiveness,) are awakened to sensuousness at an earlier period than the other faculties, and have begun to develop somewhat in advance of the faculties of Destructiveness and Cautiousness, (which are among the Propensities;) while in the other, the faculties of Destructiveness, which are those which give executive force and possess the child with the desire to act for itself and supply its own wants, and the faculties of Cautiousness, which give self-control and guidance when learning to walk, are active and developed in advance of Language. In the newly-born infant, when food is the paramount necessity, the faculty of Alimentiveness is very active, and the mother may observe the organs of this faculty bulging forth on either side of the head, in front of the ears. When the function becomes fixed, and other surrounding faculties, in turn, come into activity, this special and temporary prominence is no longer obvious. When a child is learning to walk, the predetermined necessity for Cautiousness to carry out the resolution, singularly develops the organs of this faculty, and gives them temporary prominence on either side of the head. The skull being soft, the organs appear plainly in this way, as long as the necessity for so great conscious restraint over the action of the body continues, and afterwards they retire to their proportionate size. These facts are important evidences of the truth of Phrenology.

AGE.—If the Spiritual Faculties are awakened, age is a period of strength. If not, the decline of the Propensities makes it a period of weakness.

Development of Faculties dependent on Climate, Civilization, Inheritance.

The diversities of succession in the development of special faculties or clusters of faculties depend chiefly upon conditions of climate, civilization, inheritance of parental disposition, and education.

1. *Climate.*—The climates of the middle latitudes favor intellectual development more than those of higher or lower latitudes. The warmer climates tend to give an earlier development of the Social and Animal Propensities.

2. *Civilization.*—Men naturally live in the Social and Animal Propensities; and this is the condition of the barbarian. Civilization quickens the Intellectual Faculties, making them influential and useful in guiding the forces of the Social and Animal Propensities; in truth, civilization without Christianity is incomplete, and for its permanent continuance depends on the awakening of the Spiritual faculties, which are awakened only through Christianity.

The process of mere civilization, in its mental aspect, is the process of an awakening of the Intellectual Faculties. Necessity stimulates men to more systematic labor, to more careful observation, and to the exercise of ingenuity; and hence come Arts and Sciences. Still, men do continue to live in earthly things, until the Spiritual Faculties are awakened. When this is done, the Spiritual Faculties in the individual gain the ascendancy over the Social Faculties, by their equalizing and universal force; and when this shall be understood, and *Society* thus affected, men will be led in their true social order. Civilization, in its present practical influence, affects only the Propensities and the Intellect, and it is only by the indirect influence of Christianity that it becomes fixed in the race, and continuously advances. When Christianity is properly understood, as the science of mind will make it appear, Society will progress, directly and continuously, in the proper Spiritual and Intellectual order.

In what is hereafter said of the successive development of the faculties, reference is had to civilized communities, in which the Intellectual Faculties are developed early in life.

3. *Inheritance.*—Parentage influences both the structural proportions and the relative succession in which the Social and Animal Propensities are developed; but the *activities* of the Propensities begin with the external and general sensational life, and depend on the necessities of the internal condition, and of the circumstances with which the individual is surrounded in infancy and youth. If a child has the proper structural order, and through parental carelessness and inattention is placed under the pressure of great necessities, its sensuous faculties are called into constant and vigorous exercise, and hence results an active and practical development. This is the condition of the street children in New York. Being cast upon their own resources to a great extent, the Social and Animal Propensities are called into great activity.

Influence of Public Schools. Intellectual Faculties and Propensities inheritable.

The modifying influences of our public schools are, in this respect, more important than those of any other institution. If the circumstances in which such a child is placed are favorable, the great force of character which results from the activity of these faculties is turned to the development of the Intellectual faculties and to useful employments. This is the characteristic of what are called self-made men.

From the questions which you have put to me it is evident that you have observed the mental differences existing among the children in this city, which largely arise from the variety of nationalities composing our population.

Rich men's children do not usually possess the power to acquire wealth, because their Propensities have not been brought under the necessities which result in the requisite activity; and poor men's children often do possess the power because they are brought under such necessities.

The Intellectual Faculties are strongly affected by parentage, and by caste which tend to centralize the Intellect, and to reproduce, over and over again, in successive generations the same order of Faculties, or mental disposition. This influence operates by a transmission of the form or structural development of the organs, and by reproducing the same activity. The intellectual activity which characterized the father, is more frequently inherited by the daughter, and that which characterized the mother is more frequently inherited by the son.

The heritable character of the Social and Animal propensities is more extended than that of the Intellectual Faculties, and affects the nature of the race. The characteristics of one person, in Social and Animal propensities, run through a number of generations. But the law of cross-inheritance between the sexes, which is observed in the case of the Intellectual Faculties, does not hold good in respect to the Social and Animal Propensities, because, by the design of the Creator, the Propensities have a character which relates to the sex of the person,

Education.—Education as now administered, is a process of improving chiefly the *Intellectual* Faculties, and especially the Perceptive cluster lying at the base of the frontal region, these being the faculties that are more directly awakened by the physical sensibilities, and that are exercised objectively, with matters of fact. It is therefore a partial system, addressing its chief labors to a special department of the mind.

But even this partial work is not done in harmony with a proper knowledge of the mind, but blindly and fragmentarily, and according to the opinions and idiosyncrasies of each teacher.

For instance, if the teacher has a strong verbal memory, he will labor to make his pupils good garmmarians. If he has strong Combinative and Conceptive Faculties, he will theorize with the children, and endeavor to explain to them relations of things beyond their power to understand.

Hence it results that the present application of the methods of education, without due regard to the mental dispositions of teacher and scholars, tends to develop such of the faculties as in any child may be predominant, but to give very little chance for the development of those which are not in accordance with the teacher's predominance; for, as is elsewhere remarked, the method to develop any weak faculty, is to reach it through the strongest faculty in the same group.

It is a fundamental principle in dealing with the mind to get at the primary forces first. In children, the Social Propensities are the primary forces; and the effort to awaken and train any other class of faculties is best accomplished by having an eye at all times to these prime movers.

2. *Where any Organ exists in Excess, What is the Proper Treatment?*

The excess of any faculty consists in the fact, that the sensuous activities of the mind are centered in that faculty so that it wholly leads the mind, leaving the other faculties too much behind. The problem for the teacher is to diffuse and direct this centralized activity. In order to do this, he should seek to combine and associate with it, the sensuous activity of such other faculties as will conduce to the welfare of the child; and by creating, in this way, a divergence of the centralized forces, he will prevent an undue predominance of the faculty in question, and will lead the way to the combined development of others, and prepare for any special course of training which he is called upon to give. To adapt himself and his education to do this wisely and well, is the most important function of the teacher. A leading exercise of the sensuous disposition —characteristic of children—should be the direct means of changing the natural inclination of the predominant faculties.

In regarding the excessive development of faculties in children, the first group to be considered is that of the Propensities. It must always be remembered that the attention of the child must be secured and maintained in outward and sensuous life; and through this the centralized faculties which exist predominantly in him must be guided and controlled.

The Propensities lie, as delineated on the bust, behind the ears, extending below and above them, except Alimentiveness, which stands in front of them, marked No. 1. Their general position is marked upon the bust —" *Region of the Propensities—Social and Animal.*" The physiological law by which these are governed, forms one of the most important general principles on which the education of the faculties is to be founded. This law is, that Cautiousness and Secretiveness, if predominant, exercise restraint over the action of the other faculties of the group. By restraint I do not mean a repression of activity, but a voluntary retention and economy of the vital forces. The Intellectual Faculties, being of an analytic and distributive character, without this predominance, act somewhat spasmodically, and become readily exhausted.

Influence of Cautiousness and Secretiveness, when developed in a greater or less degree.

The faculties of Cautiousness and Secretiveness, therefore, are the conditions on which we are to rely for a continuous supply, from the sensuous and temperamental disposition of the vital forces on which the Intellect depends for its continuity and economy of power. These restraining faculties are essential to conserve the forces that may have been brought out by the activity of the Propensities, and by the teacher's exercise. If these are large in proportion to other faculties in the same group, their restraining influence over the associated sensuous Animal Propensities will appear, and they will afford the teacher direct and normal means of control, at the age when full self-consciousness does not yet exist.

Too great a predominance of the mental activity of these faculties, however, producing too much restraint, checks the free, sensuous and voluntary movements of the mind, and keeps the activities within the Propensities, turning them to the worst and most vicious account; or, in case of a child of inactive temperaments, resulting in laziness and indifference, arising out of weakness and inattention from the want of sensibility.

The teacher should observe whether the restraining faculties in the Propensities predominate; then, in order to gain the attention of any Intellectual faculty, sensuous with indirect means should be applied, and the teacher must secure the good will through the Social Propensities, and thus may teach through other scholars, and by inference, rather than by direct appeals. Ignorance of the mind in this respect leads the teacher into very pernicious errors. Appeals to emulation, and temptations to dissimulation, delude the children with false ideas, and develop pride in those children in whom Self-Esteem and Approbativeness are large; while in children in whom these faculties are small, such treatment tends to encourage jealousies, quarrelling, and fighting. Other and grosser vices result from the misdirection of the vital forces which should flow to the development of the Intellect.

Children who have Cautiousness and Secretiveness small, may be very easily led by the teacher through other motives, but their minds are not retentive; they may learn readily, but lose as readily what they do learn. It is the child of large Cautiousness or Secretiveness who is not so easily led by the teacher, who seeks for knowledge for himself, and by these restraining faculties holds on and hoards what he has gained, but otherwise would lose in the current of events. The tendency of these same faculties is to retentiveness in all other mental dispositions, especially if Self-Esteem and Steadfastness be large.

If the child with whom the teacher is dealing has these restraining faculties large, the teacher has on that account more difficulty in guiding him, but has the conditions of greater success, if he can succeed in guiding him. On this disposition depends the character of self-sustained and self made men.

If Cautiousness and Secretiveness are very large, care should be taken by the teacher to draw the activity away from them, and to exercise and exhaust the forces with intellectual effort. The object as well as duty of of the educator, is not only to impart information, but to lead his sensuous nature to direct all the mental processes.

The sensuous nature of boys requires that the teacher should take advantage of every external circumstance and incident to secure their sympathy and engage their thoughts. They must be employed and never allowed to be idle, lest they form bad habits from inattention to the tasks of the school-room.

To do this the teacher must first understand their existence, and how they exist in himself. Not until then can he apply true methods; for he can properly apply only what he possesses. The law of life is movement, necessity, restraint. Intelligence is manifested by the cultivation of the Intellect, when there is restraint by the predominance of Cautiousness and Secretiveness in the Propensities; and by the orderly and proper exercise of the Spiritual disposition, comes guidance and judgment. Hence, education should be strictly a physiological and sympathetic work, not one of mechanical routine.

The particular faculties which, in school boys, are prone to be too predominant over the others, are, among the Propensities, chiefly Destructiveness, Combativeness, Adhesiveness, Secretiveness, and Cautiousness.

If *Destructiveness* is too large, set the boy to doing something useful. This is the executive faculty; and the teacher will best get possession of his mind by teaching him those things that require activity; to run, to walk, to kneel, to sing. Destructiveness always wants to be kept busy. It will keep busy, even if it is in pinching the boy next to him. Having, by some active employment, secured the attention, the teacher may call into exercise the faculty next largest to that of Destructiveness.

If *Combativeness* is too large, appeal in the same way to Cautiousness, if that be large; or if not, to Approbativeness or Self-Esteem. If Combativeness is very troublesome, isolate the boy from those he is accustomed to irritate, or put him among larger boys whom he will have to fear; and in extreme cases, overcome it by the counter irritation of chastisement.

If *Adhesiveness* is large, the boy is governed more by his comrades than by his teacher. The boy that sits next to him has more influence over him than the master. The first thing the teacher has to do is to get between him and the chum. It will be advantageous to separate him from his intimate friend. If the teacher could make a child of himself, and create a sensational attention, he could get possession of this boy's mind by means of this very faculty of Adhesiveness. The best way to get his attention is through Cautiousness, that being the higher of the restraining faculties of this group, and one which will induce him to listen. Having thus directed his attention to the subject, exercise the Intellectual Facul-

How to manage excessive Secretiveness, Cautiousness, Approbativeness or Self-Esteem.

ties. Through Adhesiveness excite Approbativeness, which is contiguous, and instead of allowing him to exert his Combativeness, excite his Self-Esteem, which is the highest faculty of the Propensities, and will give him a higher range of motives.

If *Secretiveness* is too large, the teacher may employ the boy in monitorial functions, and if Self-Esteem is also large, give him some control and direction of affairs. This will secure his attention and interest, and the teacher may then proceed to call other faculties into exercise as above stated.

If *Cautiousness* is too large, seek to influence the child through his affections. Fear will paralyze such a mind. To make this faculty useful when it is too predominant, the teacher must get the affections of the child, and he can then by proper direction make it an intelligent restraint.

The particular faculties in *girls* which are prone to be too predominant over the others, are Philoprogenitiveness, Adhesiveness, Secretiveness, Approbativeness, Self-Esteem and Inhabitiveness. Most of these faculties being among the Social affections, girls are less troublesome than boys, having less desire to assert themselves individually, and being drawn to each other by social attachments and affections. The contiguous faculties of Self-Esteem by reason of their contiguity tend to a precedence of activity in connection with the adjacent faculties of Approbativeness. The faculties characteristic of girls, therefore, afford the teacher easier means of control. This difference of development between girls and boys is by the design of the Creator. The difference in the shape of the back part of the girl's head, from that of the boy, may be readily discerned by an ordinary observer.

If *Secretiveness* is excessive, the girl acts from indirect motives, and will be prone to equivocation, falsehood, and cunning. The proper treatment will usually be to appeal to that one of the faculties above named which is the largest, so as to get the affection of the mind and obtain control.

If *Approbativeness* is too large, the child is ambitious to be distinguished beyond her proper relations. She is prone to regard more what the teacher thinks of her than what she really is or does. If Secretiveness is also large, the child is prone to dissemble, for the sake of securing the good opinion of others. The teacher should endeavor, while maintaining the affections of the child through other faculties, to call her attention continually to what she is doing, and lead her to regard more the facts of her conduct, and less the opinion of others.

If *Self-Esteem* is too large, in connection with Secretiveness, the child shows a tendency to pride, holding herself aloof from the others; and if the teacher seeks to counteract this, merely by appealing to Approbativeness, the danger is that the fault, though somewhat modified, will be confirmed. The teacher should rather endeavor to awaken Cautiousness, and call into requisition moral disapprobation, and then lead the activity into the Intellectual Faculties.

Directions when Propensities are large and active, or small and sluggish.

The Propensities have their proper order of development, which is indicated by the numbers on the bust; Alimentiveness, being No. 1. These numbers, it should be observed, relate to the organs in the adult subject, and point out the appropriate and natural order of the ultimate development in him. The teacher, therefore, should distinguish between the order proper to be followed after the age of puberty—which in some cases occurs earlier, and in others later in life—and that which obtains before that period.

3. *What in Case of Deficiency?*

The physiological force which is necessary to give vivacity to the Intellect comes from the Propensities; and the remedy for any general deficiency of activity in the Intellect of a child is to be looked for in the Social and Animal Group. If the Propensities are large and active, the teacher may draw forth the force of the desires so as to awaken the Intellect. If they are inactive and sluggish, through the weakness of temperamental conditions or through habits of life which exhaust the vital forces in bodily exercise, they must be engaged and exercised under conditions favorable to mental activity. If, though already active, they are small, so that their force is soon exhausted, they must be exercised under conditions favorable to their growth and development in size. For this purpose proper food, out-door sports, gymnastic exercises, social pleasures, and all the rough-and-tumble life which belongs to children, should be provided and encouraged. But the remedy for any deficiency in a special part of the Intellect must be looked for in the other faculties of the Intellect. Give me a child of strong Propensities, and a sensuous disposition, and the energy which he possesses may be drawn forward into the Intellect. If, however, the energy is already active in a limited part of the Intellect, the exercise of contiguous faculties must be resorted to, in order to combine and centralize, and give fulness to the activity in the deficient faculty or faculties.

If the organs of Cautiousness and Secretiveness are small, the teacher cannot maintain the continuous attention and consciousness of the child by them, but is compelled to resort to an appeal to such other faculties of this group as have a predominant development; for instance, the sentiments of Self-Esteem, inciting the subject to his own personal pride; or Approbativeness depending on the strength of praise which the teacher may give; or Adhesiveness, through which the subject is influenced by attachment, or either of the other Propensities lying contiguous to these, which may be next in predominance of development in the group. So, when activity is secured in the Perceptive cluster of the Intellectual group, the same law of operation according to predominance of development should be followed there. It is especially the duty of the teacher of young children to attend judiciously to the development of these last, as it is through them that the proper base of facts for development of all the

Intellectual Faculties is accomplished ; and the training of the Perceptive Faculties, therefore, is the foundation work of Intellectual education. The teacher must employ an influence which will awaken the predominant Propensities, and thus maintain the attention of the child. If too much restraint is exercised by compulsion and fear, the mind becomes stultified ; and if there is not sufficient restraint, the attention is not continuously secured. Moreover, nothing is more necessary in the development of the faculties of the Intellect, whether those of the Perceptive, or of the Conceptive, or Combinative cluster, than that the teacher should exercise consideration and attention to discern by which faculty or faculties it is that he has control over the child's attention and will ; whether he works by the law of the desires of the Propensities, influencing Destructiveness, etc., in males, through Alimentiveness, or Adhesiveness, etc., in females, through Philoprogenitiveness ; or whether it is by exciting the exercise of the individual's power of restraint through Cautiousness and Secretiveness by the use or fear of physical pain.

There is a characteristic difference between males and females, in the predetermined size of the organs of the Propensities. The faculties in the two sexes, differ essentially in their peculiar natural developments ; and the teacher needs to know this fact, and keep it in mind, and adapt his methods to it. The boy is overflowing with restless activity, arising from the superabundance of physical force in the organs of the Propensities at the base of the brain, and through all the temperamental and vegetative functions of the body. This must be expended somewhere ; and if the boy is properly educated, it will be expended through the Intellectual organs. Destructiveness may be appealed to, and by its proper and legitimate exercise, may call into activity the Perceptive Faculties, by drawing the attention—for example, to a fire burning, and by means of that, the faculty next contiguous to Destructivenesss, Alimentiveness, may be addressed by teaching him to cook his own food, and in turn, Secretiveness, in watching over the fire, and Cautiousness, in fearing it ; and so also with the other faculties of the group. These operations all involve the diversion of the forces of the Propensities into the Intellectual and particularly the Perceptive Faculties. And on the same principle the exhibition of fondling, caressing care, the natural expression of Philoprogenitiveness, is a special means of influence with female children. If a boy be vicious, possession and control may be got through attending to his physical life. Giving abundance of food will render sluggish the mental activity ; and continuous hard labor will draw off the vital forces from the brain to be spent in the body.

4. *In What Order should the Faculties be Trained ?*

The proper order of development in the adult is indicated in each group by the numbers upon the bust. The period of education is the period for approximation to this order. One of the most important points with children is the order of the restraining faculties with reference to the other Propensities.

In early childhood, if Cautiousness and Secretiveness are small, there is an absolute necessity, in order properly to develop the child, that he should be brought under the influence of physical pain; after which, mental fear exists, and so may be resorted to. To secure and maintain attention is the first condition for teaching. In order to maintain attention so that the instruction of the teacher may pass into the mind of the scholar, the faculties of Cautiousness or Secretiveness, or both, must be active; and, if necessary, they must be awakened by physical pain. Pain must have been *experienced*, before the child can be brought under the influence of a proper mental fear of pain. This law of punishment has many gradations of means, and is the only proper law by which the teacher can begin to regulate the Propensities in this class of children. In such, an actual physical pain must precede any successful resort to mental fear. If actual pain is not employed as a means of developing the power of restraint, the very means taken to check the child, will in many cases, tend to increase and strengthen that resolution or self-will in the child which needs to be restrained. It is a great error in many modern systems of education that the teacher endeavors to dispense with actual pain, overlooking the fact that pain and restraint are, since the fall of man, the necessary conditions of his development. Some teachers, taking this course, endeavor to substitute a mere mental fear; but this effort is necessarily futile with most children, as above shown. Other teachers condemn all appeals to fear, and rely wholly on persuasions addressed to the Propensities, thus stimulating pride and selfishness. It is an equal error, to think that the infliction of pain is the constant means of education; but many teachers think thus, and depend on physical fear too much, and stupefy the mind of the scholar. Excessive punishment by the infliction of pain, defeats its own end, by restraint which turns the force of the passions in upon themselves. If passive obedience and sycophancy were the aim of education, this would be the true means.

Moreover, it is always necessary to consider, in attending to the education of any individual, what are the functional developments in him, and at what period of life he has arrived. After puberty, a new set of motives, peculiar to manhood or womanhood, begin to manifest themselves, this being the period of sensuousness. The individual is no longer subject to restraint from Cautiousness and Secretiveness alone; but must be so directed as to bring his education under the influence of his own consciousness. If it be a vicious consciousness, it must be exhausted by discipline, and active mental and physical labor. The true preventive and remedy, however, for a deficiency in the restraining faculties after this age, is one which teachers do not sufficiently regard—*viz:*

that by the influence of the Holy Spirit there should be a change from the Animal to the Spiritual disposition. If there has been such a change in the teacher, this work is easier with the scholar. The form of religion is merely the initiating order under which the change is received; the change must be such that the subject shall have an inward realizing sense of it.

We have been speaking of the inward influence which the teacher may use; but there is also an outward one. The outward interested influence of the selfish love existing among boys, through the faculties of Adhesiveness and Approbativeness, makes them desire companionship, not only in their sports, but in their studies and labors. This love of companionship and attachment to their associates, combined with Secretiveness, will often cause them to be tempted successfully to the commission of wrongs, of which, but for the predominance of these sentiments, they would not have been guilty. When these faculties have large development and are actively exercised, it is almost impossible to come at them directly, for the purpose of controlling and restraining their pernicious results; and it is only through the good influence of other boys, with whom they are in daily companionship and for whom they entertain a warm attachment, that they can be reached and reformed. Hence such attachments must not escape the eye of the parent or teacher, but must be carefully watched and properly controlled, lest as daily experience and observation in the school-room attest, they may result in the perpetration of serious improprieties.

5. *What is the Proper Classification of the Faculties in Respect to Education?*

The faculties are to be regarded and treated as they exist—*viz:* in three groups, the Animal, the Intellectual and the Spiritual. Of these, the Intellectual group, is, with reference to intelligent education, the most important.

The position of the Intellectual organs is in the front part of the head; and is pointed out on the bust by the words, " *Region of the Intellectual —Combinative, Conceptive, and Perceptive Faculties.* " The Propensities have been already shown to be predeterminately active, under the law of inheritance of physical development up to the age of puberty and manhood, at which time the Propensities become mature, and have a fixed mental character. This character is centralized naturally according to the order of the Social group; but it is the office of religion to awaken the faculties of the Spiritual group by the instrumentality of the Holy Spirit, by which the predominance of the Propensities is superseded. During the educational process of development, the Intellect is brought into exercise, beginning in early infancy, in either cluster of the Intellectual group, as nature may direct, according to predominance of development, but generally in some of the faculties of the Perceptive cluster.

The classification which presents the faculties in the aspect of the practical administration of common education in our country may be indicated as follows.

1. THE SOCIAL AND ANIMAL FACULTIES.

1. *In Boys.*	*In Girls.*
a. Destructiveness.	a. Philoprogenitiveness.
b. Alimentiveness.	b. Adhesiveness.
c. Secretiveness.	c. Inhabitivesss.
d. Cautiousness.	d. Secretiveness.
e. Combativeness.	e. Approbativeness.
Following these, the Social Propensities mentioned in the other column.	Following these, the Individual Propensities mentioned in the other column.

II. THE INTELLECTUAL FACULTIES.

1. *Perceptive Cluster.*

    a. Language.
    b. Individuality.
    c. Eventuality.

d. The various other Perceptive faculties in their range from Individuality outward.

2. *Conceptive Cluster.*

    a. Comparison.
    b. Casuality.

3. *Combinative Cluster.*

III. THE SPIRITUAL FACULTIES.

1. Godliness.   2. Intuitive Cluster.   3. Meditative Cluster.

Since then, the Propensities require the primary attention, and a due regulation of them is the condition of training the rest of the mind, we place them first in this classification. To regulate them and use their forces, we take them as we find them, and name them in the order of their usual relative force in school children. Boys commonly have more force than girls in the individual and executive qualities, as contradistinguished from the social or domestic faculties. These demonstrative powers are with them the predominant faculties, and generally in the order above named. Girls commonly have more predominance in the domestic or home faculties. Accordingly the forms of boys' heads differ characteristically from those of girls.

The beginner in Phrenologic observation will have no better point to which to direct his attention than this; for the different development shows strikingly the different dispositions of the sexes. The girls' heads are fuller in the region of Philoprogenitiveness and Adhesiveness, compared with the general dimensions; the boys' broader in the more central part, in the region of Destructiveness, compared with general dimensions. And if a head possessing this characteristic fulness in the social cluster is found on a boy's shoulders, the observer may be confident that he has found the index of that affectionate disposition which has given rise to the term "girl-boy." The exceptional forms, both in this case and that of a "masculine girl," will confirm the general contrast. In attending to this characteristic difference, the observer may see how the same faculties consort differently in different heads, the general uniformity of girls' heads being diversified by contrasts in detail, according as one faculty or another leads in a particular instance. In some, the pointed form of marked development will be found indicating sharpness of mental activity; in others, roundness will indicate more force and fulness when aroused; in others, again, the flat form will indicate that the faculty in question spreads, giving its influence as a secondary force qualifying those which are contiguous to it.

6. How May the Perceptive Faculties be Trained?

In the first place, the teacher must recognize the existence and characteristics of this class of faculties. It is not enough, with reference to these or any other faculties, to suppose merely that there are such faculties in the mind; but, to train them properly, it is essential to individualize them, to designate them by proper names, and to discern their relative positions and size, and the spirit and order under which they manifest themselves. In this way, the teacher should learn the order of these faculties as marked on the bust, and as they exist in the teacher himself. The development of the Perceptive Faculties should generally precede that of the other Intellectual Faculties. This is the order of science itself, which begins with objective facts, and makes them the basis of reasoning, even so far as in ascertaining the conditions of mental disposition by which man is in communication with his Maker, and the manner in which the Truth is possessed.

The present popular system of education tends to give precedence to the development of the Perceptive Faculties, but, without such a discrimination between the different classes of children, as to cultivate the Conceptive and Combinative clusters, in the due order, to give strength to the Perceptive faculties.

The general law of Sensation requires the training of the mind of the child by two or three faculties and not by one alone.

The Perceptive Faculties are of the greatest importance to education, as it is through them that the knowledge of external objects is obtained. They are included in the Intellectual group of faculties; and in giving them their true order, we are instructed by science and experience that two faculties of equal force if united or associated together, must act with greater force and efficiency, than when dissociated. This physiological principle is the cause why the organs so associated together, through the middle line of the head, on the contiguous sides of the right and left hemispheres, when equally large as other faculties in the same group, have a leading influence in the character, generally. Therefore we assume that Individuality is the first Perceptive Faculty to be educated, and we have designated it as No. 1.

This is the faculty which gives clearness to the Perceptive, Conceptive and Combinative faculties, and clearness to the combination of moral ideas. It individualizes objects or phenomena, each in its singleness or oneness, in name, location or qualities, so as to separate and distinguish, in all the mental processes, the things which the Perceptive Faculties discern, or which rest in the inward consciousness. This faculty is best trained by a teacher who possesses it large, and who understands its combined activity. All the leading teachers of our schools should possess this organic condition. The teacher will find that the right use of this faculty, in his own mind, will very materially assist him. In dealing with large numbers of children, this faculty is often overtasked or confused in the effort to distinguish names and persons. The teacher should not depend upon memory alone for their names, but should have the name of each child plainly inscribed above its seat, so as to localize each in his memory, and enable him to call upon any one on the instant. By this means, the teacher standing in his proper place, would always be able to see what pupil is least inclined to give attention, and call him by name. This will facilitate the teacher's work, and bring the child more under his attention and control. The child too becomes properly seated by habit, and order is thus established. The teacher should exercise this faculty in the children, by calling it into activity in combination with the contiguous faculty which is largest in them, and in this way the attention can be retained in the Perceptive Faculties. Thus if Form is the largest of the Perceptive Faculties in the child, he should be trained in discerning various common shapes each by itself; and by constant repetition, his idea of each should be made distinct, clear, and individualized in his mind. So in combination with every other faculty of this group; and with each observation, the proper word should be taught, thus exercising also the organs of Language. It is not until an advanced stage of schooling, that the child can be be expected to have a distinct individualization of the inward consciousness. Meanwhile, he must largely be taught by memorizing, that is, by the exercise of the faculties of Eventuality and Comparison.

Under the organs of Individuality, and more interior in position, lie the associated organs of Form. When large, these organs are indicated by spreading in breadth, pressing the eyes apart from each other. It is breadth between the eyes that gives the base for all optical measurements of form and size. I mention the faculties of Form and Size next after Individuality, thus giving them precedence over other organs that are contiguous to Individuality, because they lie at the base of the brain and are immediately connected with the organs of sense, which circumstances give them a precedence in activity; and moreover, the organs of Form are contiguous to each other, and the organs of Size are contiguous to those of Language. Above this lower range of faculties, lie the organs of Weight and Locality. Those of Weight are immediately contiguous to those of Language, and this association gives them precedence, in activity, over those of Locality. The two organs of Language lie above the eye-balls and interior upon the orbiter plate. They are not contiguous to each other, but dissociated, being separated by the organs of Form. Anatomically speaking, they lie like bands, beneath and across the whole range of the Perceptive Faculties, thus connecting them all through the faculty of Language. When large, these organs of Language press forward and downward upon the eyes, making the eyes stand out, and often causing the under lid to project as a pouch.

Eventuality (in which, in connection with Comparison, are centralized the memorizing activities of the mind,) is next in order, because its organs are contiguous to each other. Locality completes the enumeration of this part of the Perceptive Faculties.

It will, however, be observed, that upon the bust, the numerical order differs from that pursued in this statement, by placing Language as second instead of fifth. The reason that I pursue this different order upon the bust is, that although some activity in the organs of Form and Size must precede a full activity in those of Language, on account of the sensuous location and relation of former organs, yet in man, as he is in civilized society, and in the practical training of children, the activity of the faculties of Language, situated as they are, leads that of the faculties of Form and Size. Words are given and used in teaching, as signs, and a necessary means of awakening those faculties and individualizing outward objects by them. When men possess the order of development in which Language is subordinated to Form, Size and Weight, they possess a superior practical intelligence on this account; but, as I have said, these are exceptional cases.

The proper time for teaching the languages is in quite early infancy, when these faculties are naturally exercised in a growing condition. Language has been commonly taught, through the ear only, as in first teaching the mother tongue, or through the eye only, as in teaching dead languages. The reason why the modern method of teaching by the black-board and by the sound, at the same time, is more successful, is because two of the senses and the sensuous organs connected directly with each, are brought into operation on opposite sides of the principal faculty, giving a double force, and a greater stimulus to the mind.

The organs of Form, Size, and Weight, are to be trained through the senses of sight and touch. The system of Object-teaching is especially useful for this purpose. Eventuality is trained by requiring the scholars to relate events, to narrate the current of affairs they see; and, as they grow older, to compose narratives in writing, and to turn their thoughts into their own minds, and see how far they can become conscious of the thoughts.

Locality is to be trained by such studies as Geography, Anatomy, etc. In these studies the pupils should be taught to realize the location. The use of globes strengthens the faculties of Locality more than maps can. Memorizing teaching is often lost. It is not by memorizing alone that the appropriate instruction for these faculties is to be retained; but involving the consciousness of the pupil in the use of the predominant faculty in this cluster, and thus exercising Locality.

7. *What Faculties May be Regarded as Conceptive?*

These also have a special locality on the bust. They are Comparison and Casuality, and lie above the Perceptive and below the Spiritual faculties.

The Conceptive faculties are those by which notions or ideas are originated or combined, thus deducing generalizations or abstractions. In this they deal chiefly with the perceptions given by the faculties of the Perceptive cluster, or with the facts of consciousness; but they are called into action not only by the Perceptive faculties, but, by all the other parts of the brain, either the Combinative, or Intuitive or Meditative cluster, or by the passions.

The science of logic, which by some has been defined as the laws of thought, is but little more than the laws of the processes of these faculties of Causality and Comparison, under the necessary conditions of perceptive facts, and in the form given through the faculty of Language. When the Perceptive faculties predominate, the characteristic process of thought is by the observation of external facts, and deductions are a secondary and minor element. When the Conceptive faculties predominate, the characteristic process is by the conception of a principle; and the appeal to external or subjective facts is chiefly subsequent and secondary in the consciousness of the mind.

8. *How should they be Addressed and Trained?*

They should be addressed according to the science of logic and the principles of pure mathematics.

The Conceptive faculties will be excited to action by the predominant group, either the Spiritual faculties, the Propensities or the Perceptive faculties. In youth the Propensities are generally the faculties that call the Conceptive cluster into action; in maturer years, the Spiritual group or Perceptive faculties commonly exercise a stronger influence in this respect.

Comparison and Causality the Faculties of reason. Their relations with other clusters.

When a child of large Conceptive faculties has reached the stage of development in which he has consciousness in these two pairs of faculties, Comparison and Causality, at the same time, he will begin to understand what he observes, either by apprehending contrasts or by seeking antecedent causes. These processes include not only the facts and principles given by the external world through the Perceptive faculties, but also those of the consciousness within.

Children of this order of mind excel others in the ability to reason clearly. These two faculties are the Faculties of reason in the strict sense of that term.

It often arises that on account of the inability of the teacher to appreciate the pupils mind, and his ignorance of the temperaments and the organic capacities of the brain, such pupils appear stupid in early years; they are always asking the question—Why is this?—and yet do not give evidence of a satisfactory knowledge of the facts which they have been taught. When the Perceptive faculties become more fully developed in the process of education, and the mind acquires the requisite individualized perceptions of outward objects, the Conceptive powers are enabled to reason from cause to effect, in all things within the range of their action, and the special intelligence of the child then appears. When Comparison predominates over Causality, by the disposition to contrast, it gives acuteness in reasoning; when Causality predominates, the mind goes to the source of every principle, or the cause of every phenomenon, whether existing in the Consciousness or outward facts.

When the Conceptive faculties are preceded by the Combinative faculties, then, as things are seen in combination in the outward world, they are seen in their combined and relative sense. Therefore in reasoning by such a mental process, breadth of view, expanded light and luminous expressions are evolved, which are not clearly or definitely apprehended by the ordinary mind, except when the mind is centralized in Ideality, giving the form of Poetry. Hence they are led off into ranges of thought without any premises, except as these premises exist in Consciousness, and therefore, the introspection of Causality and Comparison, will be in accordance with the abstraction of the peculiar mental disposition, either by the Meditative or Intuitive cluster, or by Perception.

The Conceptive faculties, especially when Causality predominates, look for Antecedents; but the highest Antecedent, the first great Cause, man is only capable of knowing through the central faculty of the Spiritual group, the eye of the Spirit, Godliness. Although the child, by reason of his undeveloped consciousness cannot fully realize this knowledge of God, he should be taught pantomimically the outward evidences of the power of God in the physical world; and this is a point of the first and most constant importance in training the Conceptive faculties, on account of the lost condition of men by nature.

Suggestions for Teachers. Proper Method of training Faculties.

In all our systems of Teaching, the cultivation of affability of manner and grace of deportment, in our view a most important branch of education, has been largely overlooked. For, the general, outward, objective forms of Religion are regarded by hypercritical minds as either unnecessary, or idolatrous in their tendency, religion being with them exclusively an internal process. But the science of mind teaches that children should be educated in the external forms of Christianity, so that in maturer years, graceful manners may harmonize with the interior conditions of Spiritual life. The age especially designed for all growth of body and mind is from infancy to maturity, hence the great necessity of teaching during this period, religious pantomimic deportment as a system of manners, which are aptly denominated minor morals. It will be seen then that the teacher, if thus properly educated in a religious life, will be enabled to manifest pleasing exterior graces in his intercourse with his pupils, he himself being a converted man, under the influence of the Holy Ghost; for, all other exhibitions of mien and deportment result merely from imitative life.

The studies most appropriate to the education of children of the Conceptive class are those, in the study of which, Causality and Comparison are the leading faculties; such as Philology, Astronomy, Natural History and Mental Science.

Children of large Conceptive faculties require to be taught, by first imparting the principles, and afterwards they will receive the facts. On the other hand, children of the Perceptive class must be taught by imparting the facts, leaving the principle to be acquired afterwards, if at all.

In reference to any cast of mind, the proper physiological method of training, is to approach the largest class of faculties in any one group, and the largest pair of faculties in that group. Where the development of the faculties is disproportionate, that is, different from that marked on the bust, the way to develop the lesser faculties is to approach the next largest contiguous pair in the same group, so as to get the attention; and when the proper attention is awakened, the lesser faculty or faculties of the same group may be called into operation, and exercised until they attain their proper development in the order as marked on the bust.

9. *What Faculties are Constructive?*

They are, Constructiveness, Calculation, Wit, Ideality, and Acquisitiveness. We have them named in their associated order, as the *Combinative Cluster*, forming a semi-circle, of which the faculty of Constructivness, on each side of the head is the centre, and so you will find it marked on the bust.

10. *What Treatment is Proper for Them?*

The faculties of this cluster, in themselves, tend to divergence and side issues, and impede Education unless they are under the teacher's control. The teacher should secure an ascendancy over the mind, and guide it by his own Consciousness of what is necessary for the child, or with pupils of an advanced age, he may rely for this purpose, to some degree, upon their own restraining faculties and thus develop their intelligent self-consciousness.

These side faculties of the Intellect are to be led and their power concentrated by the activity of those faculties lying in pairs through the longitudinal centre of the head, where the hemispheres come in contact with each other; that is to say, the faculties of the Perceptive cluster, or those of the Conceptive cluster, or the Social Propensities, or the faculties of the Meditative or Intuitive portion of the Spiritual group. The Social or Animal Propensities have a predetermined activity, in consequence of the fall of man; and they, naturally, immediately influence the Combinative Faculties, because they lie contiguous to those faculties. The Perceptive faculties too, which have a more sensuous location than the Combinative, exercise a marked influence over them. The Spiritual Faculties have less influence upon the Combinative Faculties, because they are rarely called into activity in childhood.

In consequence of the characteristics of children in this respect, much opportunity is lost by not properly regarding the activities of these Combinative Faculties in connection with the Social Faculties. Children should have more time to play; and their plays should be the teacher's personal care, in order to gain their individual affection. For the purpose of the practical exercise of the Combinative Faculties, innocent and pleasurable sports should be cultivated. Plays which involve useful results or increase the proprieties of deportment, or which imitate the constructive occupations that will be useful in after life, all those things which quicken either of the senses, the use of tools and of all the implements of childhood's sports, should be encouraged by the teacher. This interest on the part of the teacher does more to give him the confidence of the child than any thing else he can do; and thus will enable him to lead the intellect of the child.

But for the proper training, it is also necessary to treat these faculties with special reference to the influence of the other faculties of the Intellectual group, and with reference to the influence of the faculties of the Spiritual group, so far as it is possible at an early age, the latter being then dormant.

The proper method of doing this, is to seek the largest pair of faculties in the Perceptive or Conceptive cluster, or in the Meditative or Intuitive cluster, and awaken that pair in connection with the Combinative Faculties.

Thus, if in a child of preponderating Combinative Faculties, the Perceptive Faculties—that is to say, Individuality, Language, Form, Size, Weight, Locality, Eventuality, Color, Time, and Tune—are larger than the remaining parts of the Intellect, the appropriate instruction is in Orthography, Etymology, Syntax and other departments of Grammar; to

Hints for instruction when different clusters predominate.

teach him to combine words, and to form sentences; to adapt his mind to individualize his own thoughts and sensations to himself clearly, then to individualize the proper words, and then to communicate them by the aid of the rules of Grammar; to teach him how words are constructed and alphabetically put together, and the art of pronunciation; to teach him Geometry, Draughting, Metallurgy, Geography, History, Biography, Landscape and Historical Painting, and Natural History; to teach him to arrange words musically, combining them according to the regulation of the two faculties of Time and Tune, which are contiguous to each other, and so teach this order of mind the art of enunciating words musically, and to compose music, which power depends upon a constructive order of mind.

With children of this class it is important, above all, to stimulate the Conceptive range, making them look for an antecedent for every fact, so that the Conceptive cluster may be cultivated and not lose its power through the all-absorbing influence of the Combinative and Perceptive faculties.

If in a child of large Combinative Faculties, the Conceptive Faculties —that is to say, Comparison and Casuality—are larger than the other parts of the Intellect, the appropriate instruction is rather in literary composition than in the mere use of words. In music also, minds of this cast seek to compose, and their composition is more of the artistic character than of the religious or martial character. Combinative Faculties with a predominating Spiritual development tend to Sacred music, and Combinative Faculties with a predominating development of Social Propensities tend to martial and sensuous music.

Since, in children the Social Propensities predominate, the kind of music which they love is the martial and boisterous music.

When the various faculties of the mind and the modes in which they act, and the extent to which they re-act on each other, are thus clearly defined, we have plainly before us the few simple conditions, out of the combinations of which the infinite diversity of the mental dispositions of men result.

Minds of the Combinative class should also be taught mental science, practical knowledge of the three groups, and the different localities of each pair of faculties, so as to make them as conscious, practically, of their own mental organization, as they are of their different senses—sight, smell, taste, hearing—or of the fingers on their hands, the special organ of touch. When the science of mind shall have been adopted by our teachers so that it can gradually be taught in our schools as a branch of learning, in combination with the present mode of Object-teaching, all science will combine in universal laws in the order of man's mental structure as marked on the bust.

The law of Physiology teaches, that, in the mind in its natural and cultivated state, two faculties of equal force, associated together, by the contiguous position of their organs have, by their combination a precedence of activity over organs of equal size whose positions are disjoined. The capacity to retain this activity, until the mental result is attained, will depend upon their size and bulk; for the result is interiorly possessed and is not exhibited exteriorly until the organ of Destructiveness or Executiveness gives it manifestation. The degree of this retention will depend upon the influence of the teacher by applying proper mental stimulants, or upon the restraining faculties of the pupil, if the consciousness is intelligently awakened.

The term Conceptive is applied to certain faculties because they bring together and connect mental operations—this result being due to their location—but to attain predominance over the other clusters of the mind, size must be superadded to location. Here the teacher should exercise a prudent discrimination whether the child merely memorizes by his Perceptive faculties, or conceives a thought from the Conceptive cluster. If he is required to embody this thought in words, before he takes a proper form of the Conceptive, the memorizing faculties are educated only, instead of those of ideas.

If the Conceptive cluster is larger than the Perceptive, as we have before said, the minds of the majority of children being disposed to the Conceptive will be induced to ask questions without any centralized or localized reason in the faculties of Individuality, Form, Size, Locality or Eventuality. Again if the cluster of the Perceptives is so large as to absorb the nervous influence, they monopolize the mind, and supersede the the functions of the Conceptive.

Hence, it is of the first importance that no one of the three clusters of the Intellectual group should be educated at the expense of the other two, for, if the Perceptive faculties are cultivated alone, the size of the Conceptive is diminished and their power proportionally decreased; and, thus also, with the Combinative cluster.

The duty of the teacher is to endeavor to bring into activity, first, the Perceptive cluster, being the most important in mental operations, they having for their objects material proof; next, the Conceptive range, which furnishes principles with the facts of the Perceptive cluster; and then, the Combinative cluster, which gives organization to facts and principles in the order they exist in object and subject.

By an acquaintance with the laws of growth, the teacher will acquire an intelligent discrimination in the education of the different faculties. He will perceive how one cluster may be dwarfed in size and its capacities enfeebled by neglecting its cultivation, and how excessive development results from an exclusive exercise of another.

11. *At What Stage Should the Reasoning Faculties be Addressed and Exercised ?*

In its most general sense, Reasoning is a process consisting in the simultaneous activity of two or more pairs of faculties, whether they are in the Social, Spiritual, or Intellectual Group. In this sense, there is one reasoning of the Propensities, which is physical and instinctive; another reasoning of the Intellectual Faculties, which is analytic, synthetic and comparative; and another reasoning of the Spiritual Faculties, the Intuitive and Meditative, which is feelingly infused. The reason why there is no standard of correct reasoning among men—except general opinion, which is as diverse as communities are various—is, *first*, that men are ignorant of those distinctions, and reason in either way indiscriminately, and, *second*, that, in either department, men whose organic structure varies, are disposed to difference of mental operation. Thus, in Intellectual reasoning the true standard depends on the true order of Intellectual development as marked on the bust. Phrenology teaches the true standard of reasoning, by elucidating these organic conditions, and shows that diversities of reasoning will necessarily be manifested in organizations which do not conform to the true order of scientific culture in the order of the higher consciousness and enlightenment.

The process of reasoning begins, in children, when they have reached that stage in which they have the feeling of consciousness in two pairs of mental faculties at the same time. It begins in the Propensities. When, in process of time, the Propensities, by necessity, call the Intellectual Faculties into activity, Intellectual reasoning begins, and is instinctively Conceptive; but is first consciously awakened in the Perceptive Faculties; and the training of the teacher should be directed to the Perceptive Faculties. But the proper stage at which to commence *training* in Intellectual reasoning is after the age of puberty. Before that time, education is a process of preparation for clear reasoning, as above declared; and the leading pair of faculties for clearness, is Individuality. This pair of faculties, therefore, being the first in the Perceptive cluster, and, when properly developed, being the largest in the Intellectual group, is the first and most important in order to manifest reason clearly. The primary function of Individuality has relation to the external world, giving mental status to outward objects. It also is ruled by the Propensities and the Spiritual Faculties. When its activity is introverted upon the mind itself, it serves to give a clear Intellectual perception to the action of the faculties; and, if the higher consciousness is awakened, there is true self-knowledge.

If the Perceptive cluster predominates, the former function is most active, and the tendency of the mind is to depend entirely upon what appears externally—objective philosophy.

If the Spiritual faculties in the adult predominate, the latter function is more active and the tendency of the mind is to depend entirely upon individualizing the feeling of inward consciousness, either in the Propensities or in the Spiritual Faculties,—one of the conditions of the Metaphysical philosophy. These are two initiatory processes which include all reasoning.

Perceptive Faculties first developed.   How Perceptive and how Conceptive Faculties reason.

When the best organic conditions for reasoning exist, the pupil is predisposed to reason by parable; these conditions are fully developed after the age of puberty, and when the Spiritual Faculties become awakened. The order of these conditions is, first the Perceptive Faculties, then the Conceptive, then the Combinative, with the preponderating influence of the awakened Spiritual Faculties illuminating them all. The Perceptive cluster of Intellectual Faculties is the most important class to be properly educated in preparation for correct reasoning. These are the first in order of development; and truly to understand objective teaching, these faculties must predominate, for they cognize material evidence and facts as they exist, on which all reasoning must be established. Phrenology itself, the true system of reasoning, could not be made a science without having these faculties precede all others, so as to obtain a basis of objective facts. By discerning through the Perceptive Faculties, the physical conditions, the organs, their form, size, position, and the phenomena manifested, they present the objective proofs of how God deals with the individual and also with mankind. And this proof harmonizes with the subjective inward proof which rests upon Divine Revelation and inward consciousness.

The reasoning performed by the *Perceptive* Faculties consists in individualizing a phenomenon, and affixing to it the proper word or name, as a distinguishing sign. You will see, designated upon the bust, the order in which these Perceptive Faculties are related. Individuality, which takes cognizance of each phenomenon in the outer world, and separates and distinguishes it from all others, stands No. 1. in the order of development; and Language, which connects with each phenomenon recognized by Individuality a distinct and appropriate word-sign, stands No. 2. So all the Perceptive Faculties up to No. 9, stand under the order marked on the bust. 1. Individuality; 2. Language; 3. Form; 4. Size; 5. Weight; 6. Eventuality; 7. Locality; 8. Color; and 9. Order.

The reasoning performed by the *Conceptive* Faculties of Comparison and Casuality, consists in comparing phenomena, and seeking their causes. These Conceptive Faculties, marked on the bust, are 10. Comparison; 11. Casuality. Time and Tune marked 12, and 13, belong by their location between the Perceptive, Conceptive and Combinative Faculties, and when combined with either of the Perceptive, measure events by time, and beauty of expression by sound; or when combined with the Conceptive, determine and measure time in the chronological order in history; or when combined with the Combinative Faculties, give time to tune, which is the essential basis for order in the art of music.

The order of reasoning by which Phrenology became established was first Conceptive Reasoning; Dr. Gall comparing phenomena and inquiring for causes, then establishing the general facts, in the order of the Conceptive, Combinative, and Perceptive Reasoning; Dr. Spurzheim following in the same train, gathering details and constructing a classification. The reasoning performed by the *Combinative* Faculties consists in taking up what has been perceived and conceived, and combining it in the order of Intellectual synthesis and analysis. These Combinative Faculties are marked 14. Calculation : 15. Constructiveness ; 16. Mirthfulness ; 17. Ideality ; 18. Acquisitiveness.

The order of the development of the Intellectual, Spiritual, and Animal Groups, in any individual, constitutes the mental condition upon which all his processes of reasoning depend ; and when mankind shall have recognized and accepted the necessary inference from this, men will not contend about their views, but refer all difference of opinion to difference of organization, scientifically ascertained, the predominant group, the kind of education received, and the willingness or unwillingness to receive the Truth. The civilization which Christianity affords, seeks for, and irresistibly tends towards a basis of Absolute Reason ; and some progress toward this result may be seen in the present state of controversial philosophy. The basis for that Absolute Reason will be attained by the true science of mind.

The method of awakening the Intellectual Faculties is to cultivate the senses :—sight, hearing, smelling, taste, and touch.

The first law of the mental disposition to be regarded by the teacher is, that the vital forces reside in the Propensities. Hence a due consideration to strength and vigor of Intellectual character requires that the activity of these faculties should not be too early drawn away to the Intellectual part. The second law is, that these Propensities, in order to supply their inward wants, stimulate the Intellectual Faculties into activity by irritation, and doing this they are aided by external things acting through the senses ; while in striving to supply these wants, the Intellectual Faculties seek outward signs, either in action or in words, to express the inward want or condition. Hence, the teacher should secure the control of the Propensities, in order to reach and guide the activity of the Intellectual Faculties. He should do this, if possible, by engaging the affections of the child, and if not, then by compulsory means.

The action of the Intellect, as a whole is expenditive and tends to absorb the vital forces. Thus, if the Intellectual faculties demand a greater degree of Nervous action than the Propensities supply, the individual is mentally constituted for Intellectual pursuits ; physical labor becomes necessary in addition to Intellectual effort to afford healthful activity. On account of the prior activity of the Propensities of children, resulting from predisposition, attention should be directed to the differences existing among scholars, for the purpose of ascertaining which of the two groups predominate in each, so that opportunities for recreation and play may be regulated in accordance with the order predominating. For, the observant teacher will have noticed, that, in a given number of children, whilst some will become so exhausted by their sports as to render study entirely impracticable, the same amount of exercise will stimulate and prepare the minds of others for close application.

The activity of the Propensities, when guided properly, is recuperative. They involve bodily activity, which builds up the system, producing pleasurable emotions which energize the mind; and, therefore, while children are growing, it is very important that these should have proper attention and exercise. If the Intellect is overworked when the individual is young, it engrosses and takes possession of the vital forces, which at that period should be directed largely to the development of the physical structure; and thus there is produced in the brain, the condition of which, at a very early age, is properly of the lymph quality, too early a development of the fibrous state, such as prematurely disposes the subject to the nervous temperament, gives a diminutive structure to the body, and causes disease or premature decay.

The activities of the Intellect, it may be said, are either centralizing or diffusive, according to the peculiar structure of the groups. When they are centralized, it is because of the great predominance of Individuality over its associates in the Perceptive cluster, standing in their proper order. If, by a greater breadth of structure, the Combinative Faculties get precedence, the activities become diffusive.

The Intellect, in its proper order and exercise, is dependent, as well as the phenomena of the mind itself, upon the Perceptive Faculties, for a definite and intelligent aspect, and upon external objects, to arouse attention. Objects existing outside of the mind, and apprehended by the Perceptive Faculties through the impressions made by them upon the senses, apprehended through the consciousness, are an essential condition of the action and development of the Intellectual powers. The Intellectual Faculties also have within themselves a priority in activity, predetermined by nationality, Christianity, or by inheritance and other circumstances. The degree of this precedence of Intellectual activity has in each case a special adaptation to the relative sensibility of the nerves of the senses, which in turn are largely dependent upon the temperaments for their peculiarities and influence.

PUNISHMENTS.

Having now described the principal difference in mental organization as well as in temperaments, the answer to a previous question relating to the methods of discipline will be more intelligible. It has been already observed that the choice, between coercive or persuasive means does not depend upon the temperaments; but that when either is to be used, the nature and degree to be employed are to be chosen with reference to the temperament, as well as to the mental organization.

The great difficulty which teachers find in administering effective discipline is in their own ill-regulated judgments resulting from want of knowledge of the faculties of the mind, and of the temperaments; this being needed to enable them to judge of the susceptibility of the child to the punishment inflicted.

The infliction of a certain, fixed punishment for every offence of a certain character, by whomsoever committed, appears, to many teachers, a just and impartial rule; but when the organization of children is taken into account, such a method is at once, seen to be unjust and inefficient. The nature and degree of the punishment should be adapted to the mental organization of the child, as well as to temperamental conditions of susceptibility—not to the external act, but to the internal cause of the act.

The first inquiry is—What is the cause of the offence? The second is—What is the proper channel, which the peculiar organization of this child affords, for reaching that cause? This rule is *multum in parvo*.

For example, if the offence is, that the child has not learned his lesson, we must first discriminate whether the fault is indolence, or merely inaptitude for the special study. The latter should, of course, never be punished; it is to be cured by encouragement.

If *Indolence* is the fault, the first inquiry is—What is the cause of it?

If the child of the Sanguine temperament, and the indisposition to study proceeds from the volatile disposition of this temperament, and is too strong to be overcome by persuasive means, the teacher must appeal to physical pain, or to fear, which is the apprehension of pain by the remembrance of it. Ordinarily, teachers appeal to whichever of these two, pain or fear, promise the quickest result, without regarding the necessities of the child's development. Thus, if the child is fearless, the teacher whips him; but if he is timid and shy, the teacher only threatens. An understanding of the mental conditions upon which these differences depend would show the teacher, that if the Sanguine child is fearless, he should get the possession of his feelings; and if rash, it is because the restraining faculties of Cautiousness and Secretiveness are small; and if he would correct this deficiency in the child, he should threaten first, and follow it with punishment afterward. To an offender marked by this organization, let the teacher say, that to-morrow at such an hour he shall be chastised for the offence; and when the time comes, let the punishment come inexorably. Such a method awakens the necessary fear, where it did not before exist.

In some cases, if the desired attention has been secured, it will be wise, when the appointed time comes, to say to the scholar, that, as he has been assiduous meanwhile, the punishment will be, as a favor, postponed till the next day, thus prolonging the immediate stimulus of the fear; but it should never be dispensed with, when once threatened. Sooner, or later, the teacher's word must be fulfilled.

Organization and Temperament to be regarded. How to overcome Disobedience.

On the other hand, if the child is sly and cautious, it is because Secretiveness and Cautiousness are large ; and if this be so, appealing to fear, though a very easy device for the time being, increases the mental disposition which led to the fault. In such a case, it will often be well to punish the child without any premonition, and afterward tell him the reason for it. Threats would awaken fear ; but unexpected punishment will awaken the Intellect of the scholar ; for his mind will be immediately set at work to consider in himself what was the reason of his being punished, and the faculties of Cautiousness and Secretiveness will be turned to a useful exercise in making him, afterward, watchful of himself, and cautious of his conduct. If the indolence proceeds from the preponderance of these restraining faculties, Cautiousness and Secretiveness, which is sometimes the case in a child of the Nervous temperament, the difficulty to be met is that the mind has too much restraint, begetting indifference. In such cases, the teacher must, if possible, avoid appealing to fear. He should seek to ascertain which of these two faculties predominates. If Cautiousness predominates, pain or fear will tend to stultify the mind. He should endeavor to awaken the forces of other Propensities, and then to lead them forth into the Intellect. Persuasive measures, and means, such as draw the motive power and force more into combination with other faculties, should be used, so as to overcome Secretiveness. These measures will be particularly appropriate if Secretiveness is the larger faculty.

If the cause of the indolence is the predominance of the Bilious temperament, the teacher should counteract this by leading the child to active, out-of-door sports. Those exercises which rouse and quicken the system, and give more play to the lungs and more vigor to the stomach, with rest to the brain, will be necessary to correct an indolent disposition resulting from the predominance of the Bilious temperament. If the cause of the indolence is the over-indulgence of the Lymphatic system, the teacher must take measures to have the supplies of food diminished, so far as necessary to give mental activity during school hours, and superadded to this, the discipline recommended above in cases of the predominance of the Bilious temperament. So long as the tasks given to the stomach require the nervous force of the child, it will be neither easy nor wholesome to call that nervous force away into the brain by punishment.

Wilful disobedience and insubordination will be found to depend on mental conditions to a greater extent than does indolence. The first step towards overcoming this evil is to ascertain where the will of the pupil resides ;—What are the ruling Propensities that make this disposition? It will be of the first importance, in cases of this kind, to act in harmony with the parents ; for if the child is sustained at home in his wilful purposes, the teacher will rarely be able to overcome them at school.

This disposition will be more frequently found to make difficulty in the case of Sanguine or Nervous temperaments or the Sanguine-bilious, or

Nervous-bilious. In these cases, the teacher will often be able to do more, indirectly, and through the influence of other pupils, than directly, by his own administering of discipline.

In this class of offences, the teacher should keep always before him, that the welfare of the child is at stake, and that what he has to do is to reclaim. He has a battle to fight and must use strategic measures. He must use his own faculties of Cautiousness and Secretiveness, tempered by that higher and holier influence, a loving spirit; for the disposition with which he has to deal often involves, to a greater or less degree, the minds of a number of the scholars, and to meet it successfully he must have some understanding of the extent of the disaffection and the nature of the plot. If he is sufficiently circumspect to define these to himself, and sufficiently deliberate in his discipline to wait until he can lay his hand upon the ringleader, the chastisement of that one will often gain at once the whole victory.

Children brought together form a community characterized by a great activity of the Propensities, but volatile and easily moulded.

Quarrels and ill conduct towards each other, among the children, compose a class of faults different from insubordination. Here is the training ground for self-control, and all those manly and womanly social qualities which fit the possessors for happy and useful places in society.

The teacher who rightly regards the Social Propensities, which rule the children in the play-ground, will endeavor to have the children gain the right development by the sympathies and hostilities engendered in their preparatory society; and will therefore endeavor to cultivate that *esprit de corps*, which will make the aggregation of children more closely resemble an organized community, and to maintain within it a public opinion in favor of the exercise of the moral qualities of the Spiritual Faculties. For this purpose, the teacher, while he compels order and protects against ill usage, should rely upon the scholars themselves, not only to encourage right conduct, but also to correct offences, as far as it is possible to do so, himself interfering directly, only when exigencies require it, and encouraging the children to sustain themselves and each other, against aggression and injustice, by such measures of self-protection as may be proper.

The teacher must study to ascertain whether the spirit of the school harmonizes with his own wishes, or is antagonistic. In the former case, he will find but little difficulty in its management and control; whilst in the latter, a special adaptation of his own characteristics, both mentally and temperamentally, to those of the boys, will be required for their proper government. In other words, he must be, "a boy among boys," and if necessary their leader.

EDUCATIONAL "SYSTEMS."—Most of the special "Systems" of Education which have sprung up during the past century, but soon have become merged in the general progress of Education, have had their origin in more or less distinct appreciation of some one of the foregoing principles, and in an attempt to administer education upon that one, without due reference to the others. Hence each of these systems, though put forward by its founders as an universal system, has been in fact special, and has proved to be so by failing to supersede all others, and resulting merely in contributing a new phase to some educational processes, and promoting the advancement of some departments. Thus, for instance, the Lancasterian or monitorial system had its foundation in the temperamental and sensuous affinities which are naturally stronger between children, than between teacher and child. The secret of its temporary precedence as a system, was the striking result manifested when this principle was involved on a large scale. But it was found by experience that the mental superiority of adult instructors more than compensated for the apparent advantages secured by relying on the mutual plan, as a substitute for the immediate supervision of skilled teachers. The intelligent instructor will observe the strength of sensuous influence of children on each other, which was the secret of this method; and will avail himself of that influence, constantly, in support of his own.

The Pestalozzian system was founded on a very clear and strong apprehension of the importance of educating the Perceptive faculties. Every thing was bent to this standard, and the Conceptive powers disregarded. What is called Object-teaching is prone to the same error.

Many philosophic teachers in American schools go to the other extreme, and deal with the Conceptive powers to the constant neglect of the Perceptive.

The Kindergarten, and the excessive use of calisthenics and gymnastics, or military drill, are other examples of special systems, which should not be made a procrustean bed for all pupils. These methods should all be administered with due reference to the temperamental and mental character of the pupil to be educated, so that the special advantages of each may be utilized.

Any general and arbitrary application of either system, without regard to the temperamental adaptations, is pernicious and will fail of a general success.

12. *What Moral Faculties Claim an Early Attention?*

In a strict sense, the only moral faculties are the Spiritual Faculties. The Propensities are essentially selfish in their nature; and the Intellect is merely the instrument of either.

Suggestions for Moral Training. Why Religion is generally distasteful to Young Men.

The Spiritual Faculties should not be awakened until after the age of puberty. If they are called into activity and take precedence at an earlier age, the vital forces are turned away from the proper channels for development of the body, and the growth is checked, and weakness and a cessation of bodily development result. For physiological reasons, therefore, these faculties should not have predominant control, until that age.

But there is an important part of early education which may properly be regarded as moral training, because it is preparatory to this. *First*, the teacher should give early attention to instruction in the manner, postures, language, and musical tones, which constitute the outward, objective forms of Spiritual religion, and these should be the earliest means used for directing the attention to religion. These forms do not constitute religion; but they will make the children graceful in their mien, and imbue them with a respect, in outward relations, towards religious teachings, and towards teachers, parents, and all persons who are appointed to govern. *Second*, the teacher should seek, by a right development of the Propensities, and by exercising the vital forces in the Intellect or in bodily exertion, to train the child in right habits and in innocent activities, and to predispose the mind to virtue; remembering, however, that the motives in a child's mind upon which the teacher must mainly rely are selfish motives; personal attachment, Fear, Self-Esteem, the Love of praise or rewards, and the like; and that conscious, indwelling virtue is not attained until the age at which the child is prepared to take responsibility upon himself.

The ordinary methods of religious instruction fail of general success, because of not recognizing these facts. Children are required to fast, and to conform to spiritual exercises, and are sometimes stimulated to premature experiences of moral sentiment, although the ordained necessity, even in the fallen state of man, requires the period between infancy and puberty for bodily growth, and for the development and regulation of the Propensities, and, through them, of the Intellectual faculties, as preparatory to the awakening of the Spiritual disposition. It is not strange, after such forced training, that when the age of religious development arrives, religion appears to the young man as a thing of trouble and constraint; and he avoids it, and, when too hardly pressed, cherishes dislike towards it, instead of seeking and attaining the power of a divine life, in which alone the true harmony and energy of man's nature will ever be developed.

Children may be taught, intellectually, the forms of religion by means of Catechisms, Bible records and Scriptural truths, but their Spiritual faculties being usually dormant, until after the age of puberty, they cannot readily be taught to realize fully their Spiritual nature. In the language of the Scriptures, they "must be born again;" and this results from the natural fallen state of man.

When the child becomes self-conscious, and realizes that there is a God, and that He is the highest of all—and at a more matured period, when the inward Spiritual consciousness is specially awakened by the influence of the Holy Ghost, through the instrumentality of the faculties of Godliness, which are the centre of the Spiritual group, then he will begin to show a proper regard to all things, having been trained to respect the exterior order and conditions through and by which the Spirit of God is manifested. The objective form of religious teaching which is established in the world as a preparation for religion, is infinitely more important than those other necessary pantomimic exercises, such as calisthenics, which give expression only to physical and sensuous ideas.

The true science of the mind, as well as Divine Revelation, teach that there is a universal necessity in man's nature, (which all experience confirms,) for a change from activity and consciousness in the physical and Intellectual part alone, to activity and consciousness in the Spiritual part also. This change is what Christ described to Nicodemus as being born again. The power to know God, as He is revealed by the Holy Ghost, is exercised through the Spiritual Faculties, and to know Him properly, they should stand in the order marked upon the bust:—1. Godliness; 2. Brotherly-Kindness; 3. Steadfastness; 4. Righteousness; 5. Hopefulness; 6. Spiritual Insight; 7. Aptitude. This should never be lost sight of by teacher or pupil. But this truth, however clearly it might be stated, will be of little avail to the reader, unless realized in his inward consciousness. The Intellect can only objectively comprehend it. A description in words, of a good dinner, will not feed a hungry man; and a statement of the true order and life of the Spiritual Faculties cannot satisfy the hungry and thirsty soul, in those qualities of the Spirit spoken of above.

13. *How to be Trained?*

The moral faculties, being identical as above explained, with the Spiritual Faculties, can be trained only by those teachers who are influenced by the Holy Ghost and whose Spiritual Faculties become predominant in activity through His power. As a teacher cannot train the Intellect, without realizing, objectively, the facts which he logically knows, so he cannot train the Spiritual Faculties in truths, which he does not himself subjectively and spiritually know. The inward consciousness bears the same relation to moral training, that external facts do to Intellectual training.

Intellectual things are practically and really understood only objectively and logically, and Spiritual things are realized only subjectively, by the instrumentality of the Holy Spirit. The science of mind will elucidate and define the distinction and the limits between these two regions of the mind.

In all effort for moral influence upon others, the teacher must himself possess the spirit which he would inculcate. The Intellect may make disquisitions upon moral and spiritual subjects; and remotely, by the logical presentation of divine truth, may awaken activity of the Spiritual

Faculties in others; but it cannot directly reproduce in others, that inward, conscious realization which belongs to the teaching of the Spiritual Faculties. Hence result the indifference and coldness prevalent on the most important topics of our instruction in this life.

The Intellect may be taught the organic existence of the faculties of Godlienss and the other Spiritual Faculties, but the true and proper understanding of this group can only be gained by the possession of activity in them, which is given by the Holy Ghost. The Intellect cannot have a realizing sense of Spiritual things, any more than the ears can see, or the eyes can hear.

The teacher can only acquire the self-knowledge which he needs, by adding to the objective study of the mind, that Spiritual Wisdom which the Holy Spirit offers to all who will seek it.

The Propensities cannot give us knowledge of ourselves. The passions can try only the passions. The Intellect cannot do so, it can only analyze the consciousness which is exercised by the faculties, and perceive and compare the objective forms of manifestations. This it can do both of the Animal and Social part and of itself, and of the Spirit, according to the law of each and the condition of organization; thus it may assist and describe the results of self-examination, but it has no Spiritual consciousness and cannot institute it. A true knowledge of ourselves is instituted by the Holy Spirit in the consciousness of the Spiritual fact subjectively, the Intellect may give only an objective knowledge of organization, and this is Phrenology.

The truths of Spiritual life, therefore, are of the first importance to the Teacher.

SPIRITUAL LIFE.—The objective physical facts of man's organization explain and corroborate the principles of the Gospel of Christ, and all the teachings of Scripture respecting man's moral nature.

Those mental states and dispositions which Christ pronounced blessed, are those which come, either from the predominance of what we have delineated as the Spiritual Faculties, or from that disappointment of the Propensities which is often the necessary condition of the latter being brought into subordination, by the Holy Ghost awakening the consciousness of the soul into the Spiritual Faculties. The poor in spirit, the meek, those that hunger and thirst after righteousness, the merciful, the pure of heart, the peace-makers,—all these types of character are, in mental analysis, found to be characterized by predominant activity in the seven-fold Spiritual gifts. The poor, the mourning, the hungry, the weeping, the persecuted,—all these are found to be characterized by that deprivation of the immediate gratification of the Propensities, which, we have seen, is very commonly the physiological condition of the awakening of the Spiritual Faculties, when the Propensities have long

Natural Manifestations of the Propensities.

been uncontrolled. They who are fitly described as the full, the rich, they that Laugh, and they of whom men speak well, are the type of those in whom the Propensities, Social and Animal predominate, judiciously ruling the Intellect and superseding the Spiritual Faculties.

The things which our Lord and Saviour Jesus Christ reproved, are those which are the common manifestations of the Propensities, when they are not energized, regulated and illumined by the Holy Ghost, through Godliness, Brotherly-Kindness, Righteousness, and the other faculties of the Spiritual Group. Anger without a cause, which Christ declared to be under condemnation equally with murder, comes from the same faculties whence murder proceeds, *viz:*—Combativeness and Destructiveness. That lust of the eye, which he pronounced essentially the same as adultery, is the manifestation, through the sensuous organs of sight and touch, of the activity of the same faculty of Amativeness, which gives rise to adulteries.

Not only the falsification of oaths, but swearing and evil speech, come from the Propensities, through the sensuous organ of Language, acting without restraint either by the higher Propensities or the awakened Spiritual Faculties, therefore the Scriptures declare this disposition to be set on fire of hell. The desire for revenge, an eye for eye, and a tooth for a tooth, the resisting of evil, the withholding benevolence, the not doing to others as we would that they should do to us, the hatred of enemies,—all these are the natural manifestations of the Propensities asserting themselves, in various combinations, and with the aid of the Intellect.

Love to them who love us, and lending to those of whom we hope to receive,—these dispositions, although accounted as virtues, are only the virtues of the Propensities; and however useful and desirable these dispositions are, as compared with some other manifestations of the Propensities, they appear, by the analysis of the mind, to be, as our Lord described them, essentially selfish, and wholly different in respect to their relation to other faculties, from that Divine Love which he manifested, which we see resides in the faculty of Brotherly-Kindness, and which he taught we must possess if we would be the children of our Father which is in Heaven.

Doing alms before men, and praying and fasting in public places to be seen of them, are the promptings of Approbativeness, not of Brotherly-Kindness and Godliness. Laboring to lay up treasure upon earth rather than in heaven, choosing the service of Mammon rather than that of God, and being filled with care for the things of this life, are the activities of Acquisitiveness, Secretiveness and Cautiousness, predominating over Steadfastness and Righteousness. While upon the other hand, seeking first the kingdom of God and his righteousness, and trusting in Him that all these things shall be added, is the characteristic manifestation of the predominance of Godliness, Steadfastness, Righteousness and Hopefulness, under the influence of the Holy Ghost.

Manifestations of Spiritual Faculties; of the Intellect when ruled by the Propensities.

The disposition to forgive and to judge not, is the activity of the Spiritual Faculties led by Brotherly-Kindness; while it is Self-Esteem, or a selfish will centralized in the Propensities, that causes us not to see the beam in our own eye while beholding the mote in a brother's eye.

To ask, to seek, to knock, in the belief that our Father which is in heaven will give good things, and even the Holy Spirit, to them that ask Him, is the exercise of the Spiritual Faculties led by Godliness.

To enable us to beware of false prophets, to know men by their fruits, and to discern between the good and evil treasure of men's hearts, is the function of the faculties of Spiritual Insight and Righteousness.

To hear the truth and do it not, is the state in which the Propensities refuse to yield to the Spiritual Faculties, and the restraining and executive faculties, instead of becoming the servants of the Spiritual Group, overrule them, and carry out the behests of the Social and Animal nature.

When these things are said to be hid from the wise and prudent, and revealed unto babes, we see, by referring to the actual organization of man, the physical, objective conditions of this blindness of the Spiritual Faculties, arising from the predominance of an Intellect ruled by the Propensities, in consequence of the fall of man. All the evil things which come from within and defile the man—evil thoughts, adulteries, fornications, murders, thefts, false witness, covetousness, wickedness, deceit, lasciviousness, an evil eye, blasphemy, pride, foolishness,—are the manifestations of the Propensities and the Intellect, when the Spiritual Faculties are not in control.

Not only in the didactic teachings of our Saviour, but in His personal intercourse with the disciples and with men at large, the things which He reproved were the manifestations of predominant Propensities, and those which He commended and rewarded with blessings, were the result of activity of the Spiritual Faculties.

When, to the inquiry,—" Whom say ye that I am ?"—Peter answered —" Thou are the Christ, the Son of the living God,"—He replied— " Blessed art thou." * * * "for flesh and blood hath not revealed it unto thee, but My Father which is in Heaven;" and upon this rock,—the inward conscious revelation of God by the indwelling of the Holy Spirit in the hearts of men,—He declared He would build His Church.

When Peter began to remonstrate with Him for His willingness to go to Jerusalem, to suffer death for men, He rebuked the disciple in whom thus, the Social affections were asserting predominance over Brotherly-Kindness and Steadfastness, saying, " thou savorest not the things that be of God, but those that be of men." And calling the people together, He said to them, " If any man will come after me, let him deny himself;" adding that those who should be ashamed of Him and His words, in that evil generation, should be rejected in the judgment. The fear and shame of which He thus warned them result from the predominance of the higher faculties of the Propensities, Cautiousness and Approbativeness, over Godliness and its associated faculties in the Spiritual Group.

When the people sought Him because they had been fed by His miracles, He reproved them, because it was merely the gratification of a Propensity, which led them to follow Him, and He bade them labor not for the meat which perisheth, but for that which endureth unto everlasting life, which He should give them; telling them that they must receive Him as the bread of life, and that no man could come to Him except the Father draw him.

And when He predicted His rejection and suffering by that generation, He attributed it to the fact that men were engrossed in the Propensities, eating, drinking, marrying and giving in marriage, until their sudden end should come.

Faith, which is the generic name given by the Scriptures to the childlike reliance and receptivity which characterize the mind when the passions are reduced and the Spiritual Faculties predominate, He always treated as the condition of spiritual and miraculous power; and He declared that all things are possible to him that believeth. The inability of the disciples to work the miracle they attempted, He attributed to the want of this faith, and declared the necessity for prayer and fasting.

The dispute as to which of them should be greatest, and the request on behalf of two, that they should be permitted to sit on His right and left hand, in His kingdom, disclose the activity of Social Propensities. His reply that he who should humble himself as a little child should be the greatest, that whosoever would be great among them should be their minister, even as He came, not to be ministered unto, but to minister, and to give His life a ransom for many—shows us the subordination in which all the Social faculties must be kept to the faculty of Brotherly-Kindness, in its proper order after Godliness, in the Spiritual Group.

The condition of inheriting eternal life, is to "love the Lord thy God with all thy heart, and with all thy soul, and with all thy strength, and with all thy mind, and thy neighbor as thyself."

When the rich young ruler, who asked what he should do to inherit eternal life, said that he had kept all the commandments from his youth up—Christ put the test of the complete subordination of the Propensities to the Spiritual Faculties, by calling him for a disciple, and bidding him to sell all that he had and give to the poor, and take up his cross and follow Him. The young man went away sorrowful, for he had great possessions. Upon which Jesus said to his disciples: "How hard it is for them which trust in riches to enter into the kingdom of God." Where a man has great wealth, the disposition to hold and hoard it, and to trust in it, comes from the predominance of the Intellectual Faculty of Acquisitiveness, and all the restraining faculties led by Secretiveness and Cautiousness; and if the Propensities predominate in the mind, and are led by these faculties of Cautiousness and Secretiveness, with Self-Esteem, the peculiar exercise of the latter gives great pertinacity to the worldly spirit.

The widow whom Christ commended because she cast into the treasury two mites, which were all that she had, indicated by that act a more complete subordination of the Propensities to the Spiritual life than the rich, who of their abundance cast in much.

The tests, which He describes as those by which human conduct is to be judged in the last day, are those which depend on the predominance of Godliness and Brotherly-Kindness. Feeding the hungry, receiving the stranger, clothing the naked, visiting the sick and the prisoner—these acts towards the least of our fellow creatures is accepted by God the judge, as done to Himself.

The formal observance of religious ceremonies by those who omit the weightier matters of the law, judgment, mercy and faith, the desire of chief seats and public greetings, the laying of burdens upon others which one will not himself bear—these are all manifestations of the Propensities, led by Approbativeness and Self-Esteem, and ruling the mind.

The disciples who asked Him to call down fire from heaven to revenge the inhospitality of the villagers, the persons "which trusted in themselves that they were righteous and despised others," the money changers and traffickers whom He cast out of the temple, all these are instances of hearts ruled by the Propensities.

All the things which He inculcated upon His disciples, in His last conversation with them before his death, are comprised in the predominance of the Spiritual Faculties. The humility He taught by the example of washing of feet; the new commandment that they love one another; the prayer for Peter that his faith fail not; the calming of their trouble and fear by appealing to Faith and Hope; instructing them to abide in Him and to keep His commandments; all these consist in the predominant exercise of the Spiritual Group.

To guard against misapprehension, we should distinctly observe that Christ does not teach asceticism. He teaches that the Propensities must be under the Spiritual guidance, not that they are to be suppressed. He does not teach that the exercise of the Propensities is sinful; on the contrary he promises their gratification to those in whom the Spiritual faculties predominates. "Seek ye first the Kingdom, and all these things shall be added unto you." It is the complete and perfect subordination of the Propensities and the Intellect to Faith, that is necessary; and under this subordination, both the Intellect and the Propensities are to be actively exercised, and are to find their true and highest gratification.

Christ, taught, too, that the restraining faculties of the Propensities, viz., Secretiveness and Cautiousness, duly subordinated to the Spiritual Faculties, should be exercised for proper restraint upon the activities of the other Propensities, "Be ye wise as Serpents;" "Take ye heed, watch and pray; "and what I say unto you I say unto all, Watch;"

Self-abasement. Principles established by Holy Scriptures.

"and take heed to yourselves lest at any time your hearts be overcharged with surfeiting and drunkenness, and cares of this life." And at another time he said to Peter, after he had predicted his denial, "Watch and pray that ye enter not into temptation, the spirit indeed is willing but the flesh is weak."

To be wise as a serpent and harmless as a dove, the mind must possess that watchful prudence which comes from activity among the Propensities, of Secretiveness and Cautiousness; but while these faculties lead in the lower part of the brain, the Spiritual Group, Godliness preceding all, must predominate over them and all the rest; and thus Godliness gives the meekness and harmlessness of the dove, while Secretiveness gives the watchful spirit of the mind. Thus it becomes the mind to be watchful and meek.

Self-abasement results from self-examination. Conscious self-examination acts with reference to the external and internal experience of sensuous life; hence the utter inability of the mind, unaided, to sustain itself in the belief of a future state by reason of the diversified changes of all its conditions from the beginning of life to its close.

Self-abasement, uninfluenced by the Holy Ghost, by reason of its meek and passive conditions, can only produce a state of utter hopelessness, even when aided by all the knowledge attainable by Intellectual facts, and all that physical nature furnishes in sensuous, material conscious evidence, together with the sentimental dependence, which the moral attributes impart. Hence, to possess an unfailing support in our present state of existence, and one which will bear us safely into that future which our consciousness realizes, God's Holy Spirit must be possessed; and this Spirit alone can be the life of the soul. The spirit when thus possessed manifests itself phenomenally through the physical organization.

The whole teaching of Scripture sustains and enforces these general principles, that mankind are naturally under the predominant control of the Propensities and the Sensuous faculties of the Intellect, and that this state is the carnal heart, which does not know God; that eternal life is to know God, which is through the predominance of the Spiritual Faculties, by the manifestation of Christ and the instrumentality of the Holy Spirit; that there may be a proper activity of special Faculties in the Propensities, giving morality of external conduct in certain respects, even where the Spiritual Faculties are subordinated, but that such morality is, in its nature and origin, of the flesh, and that, to bring the soul into the true relation to God, the whole of the Propensities must be subordinated by the Holy Spirit, the man being thus changed, quickened, regenerated.

14. *What other Considerations have Reference to this Point in Such a General Summary as the Above?*

The third general law which should form a controlling principle in education, is manifested in the Meditative and Intuitive, Spiritual Faculties.

Here we must assume, for a correct understanding of this subject, a new nomenclature for the faculties of this region. Dr. Spurzheim undertook to give a general classification of the faculties, and the general laws governing them. He divided all the functions of man which take place with consciousness into two orders, designating them the Affective, and Intellectual Faculties. The Affective Faculties he subdivided into Propensities, or those powers which produce only desires, inclinations, or instincts; and sentiments, which have something superadded to inclination. The Intellectual Faculties, he subdivided into the Perceptive Faculties, including the functions of the external senses and voluntary motion,—those faculties which make man acquainted with external objects and their physical qualities,—and the functions connected with the knowledge of relation between objects, or their qualities; and the Reflective Faculties, which include all those which act on other sensations and notions. The following is Dr. Spurheim's exact classsification:

AFFECTIVE FACULTIES, OR FEELINGS.

1. PROPENSITIES.

* Desire to live.
1. Destructiveness.
3. Philoprogenitiveness.
5. Inhabitiveness.
7. Secretiveness.

* Alimentiveness.
2. Amativeness.
4. Adhesiveness.
6. Combativeness.
8. Acquisitiveness.

9. Constructiveness.

II. SENTIMENTS.

10. Cautiousness.
12. Self-Esteem.
11. Reverence.
16. Conscientiousness.
18. Marvelousness.
20. Mirthfulness.

11. Approbativeness.
13. Benevolence.
15. Firmness.
17. Hope.
19. Ideality.
21. Imitation.

INTELLECTUAL FACULTIES.

I. PERCEPTIVE,

22. Individuality.
24. Size.
26. Coloring.
48. Order.
30. Eventuality.
32. Tune.

23. Configuration.
25. Weight or Resistance.
27. Locality.
29. Calculation.
31. Time.
33. Language.

II. REFLECTIVE.

34. Comparison.

35. Causality.

Defects of Classification. Error as to nature of Spiritual Faculties. "Blind Sentiments."

In this classification Dr. Spurzheim has disregarded some of the most important phenomenal aspects of the mind, constructing a theory from his own peculiar point of view and and according to his own mental organization; and, although his theory conforms to Phrenology, abstractly considered, it does not embrace all the facts exhibited in the structure and activities of the mind, and in the history of the mental life of mankind. A true classification must be based upon a consideration of the whole mental organization and all the phenomena of mental life. The history of religion must be considered, as well as the course of Intellectual development; and the Spiritual Faculties must be examined, not alone by the Intellect, which can only observe their structural order and their phenomenal aspect, but also by the inward consciousness of the Spiritual Faculties themselves, by which their true relations can be subjectively realized when influenced and orderly exercised by the power of the Holy Ghost.

But Dr. Spurzheim did not properly recognize the new birth, which the religion of Christ Jesus has shown to be of more importance, in the development of man, even than the natural birth. Hence, he was led to form a classification of the faculties by an Intellectual process—a classification which rested upon and expressed his logical and philosophic idea of their nature and action, instead of corresponding truly to the actual grouping under which they exist, and which has been pointed out previously. He was, in some measure, compelled by his own consciousness, to recognize and assent to the Spiritual laws to which man is subject, and conform his statements to the natural laws of morality. But his classification and nomenclature fail to present adequately the great and fundamental doctrine of the subordination of the faculties of man to the spiritual influence of the Spiritual group. This defect in his view, necessarily, threw the whole subject into the abyss of polemic discussion; for his philosophy violated the consciousness of men, who, not being able to discern the defect, and correct it, could not receive the view presented by him as a correct statement of the facts of our organization.

From knowledge acquired through the Intellectual faculties alone, he was enabled to discriminate between a Spiritually-minded and a purely Intellectual man. But, as he regarded all Spiritual manifestations as chimerical or visionary, destitute of logical foundation or practical demonstration, he asserted, that the world had been subjugated by Priestcraft, fanaticism, and superstition. Whilst he was forced to acknowledge the existence of the Spiritual faculties, which he denominated "Blind Sentiments," attributing to them in common with other organs, form, dimension, location and also negative moral powers, yet being destitute of the spirits illuminating grace, was incapable of portraying the reflecting, elevating and restraining influence they exert over their possessor, when enlightened and controlled by the power of the Holy Ghost. He gave them a description based upon the processes of the Intellectual faculties

Necessity of correct Nomenclature. True nature of Spiritual Faculties must be recognized.

merely, and made them the arbitrary test in man's natural fallen state. Though the Intellect does take cognizance of their structural order and phenomenal aspect, yet no one, without the knowledge subjectively given by the teachings of the Holy Ghost, can delineate their powers and capacities. Their true relations and influences must be experienced and realized, before an accurate and truthful representation can be given.

It is of the utmost importance to the success of all educational processes, that we have, at the outset a correct nomenclature—an accurate name for each faculty. This, in the present stage of our knowledge of the mind, is a task of too much importance and difficulty to be treated in a series of brief letters like these. I am obliged to leave this task to others, accepting and using for the present, the names suggested by Drs. Gall and Spurzheim, for the faculties of the Animal and Intellectual groups; arranging them, however, under a proper and definite classification in groups. My care and attention have more especially been drawn to the faculties of the Spiritual group, and to the necessity of framing a more expressive and accurate nomenclature for them, than has heretofore been in use among Phrenologists. It has seemed to me to be especially necessary, before I could begin to make clear the errors of Phrenologists, in the description they have attempted to give of the grouping of the faculties, that I should give a nomenclature of the Spiritual Faculties expressive of their nature and existence, which Phrenologists have not understood. Instead of perceiving the fact that the faculties exist in three groups, which in their proper order and individualized operations are independent each of the other, just as the senses are independent, though their operations coalesce with each other, they have treated them by Intellectual discrimination merely; and thus have been misled, by a mere objective philosophy, to follow an arrangement of classes which, though as far as it goes is substantial, is yet artifical.

If Dr. Spurzheim had lived long enough to carry out his anatomical investigations by pathological proof, and had rightly regarded the history of man in his civilized and christianized state, he would have been forced to modify his classification, by recognizing the Spiritual Faculties (which he termed blind), as being the proper ones to lead and direct. Instead of dealing with the abstract and artistic relations merely, it is necessary to review the history of mankind; and to adopt a proper classification to present the phenomenal life of the Spiritual nature, in accordance with that history. It will be acknowledged by all, that history shows that the Spiritual Faculties have had the supreme control. The influence of religious opinions and feelings has been permanent; all else has been variable and fleeting. Yet this fact has never been sufficiently recognized by those who have treated of mental science.

Religion, as now properly understood, subsists in intellectual operations, and is made to depend upon logical deductions. Too often the heart, that is the Meditative and Intuitive functions of the Spiritual Faculties, is not affected or sought to be affected. This is the great obstacle to the increase of true Vital Piety in the world. While our Colleges and Seminaries of Learning are thus confined largely to teaching Religion by the Intellect alone, their classes will abound in Spiritual ignorance and indifference, if not in infidelity. Whilst the theory of religion, the existence of God—his creative power—his superintending control is recognized and acknowledged chiefly as but legitimate deductions of mental operations—the heart remains unaffected; and that realizing appreciative feeling of God's tender mercy and loving-kindness, and corresponding gratitude on the part of the creatures of his bounty and the objects of his care may never be felt, or if felt, the proper manifestation and expression may be suppressed by intellectual dissent from the dogmatic form of the deductions.

Therefore it is of the utmost importance that we individualize this group of faculties, and locate each of them accurately, and that we correctly understand their relations to one another, and to each of the other groups, the Intellectual and Animal, and apply to them a nomenclature which expresses these facts. I have therefore resorted to the Holy Bible, as the sole guide by which the true nomenclature of the Spiritual group and its order is made manifest.

These are sufficient reasons for assuming a nomenclature, such as is marked upon the bust; which is framed especially as a Scriptural and Christian nomenclature adapted to express the order and nature of the Spiritual Faculties.

It is to the systems propounded by Drs. Gall and Spurheim, that we are indebted for the foundation of Phrenology, the one presenting the significance of peculiar prominences in the general conformation of the whole brain, and the other defining and delineating in detail the special organs, without giving the special form sufficient significance. In these letters I have endeavored briefly to delineate the grouping of the faculties, and the independent yet associated action of the groups, recognizing, too, the influence of the Holy Spirit through the Spiritual group. I have also attempted to show how the brain, as an organ, depends for its quality, and to some extent for its activity, from infancy to manhood, upon the constituent elements of the three vital functions, by which, respectively, it receives more warmth, more plasticity or ductility, or more support from the substantial physiologic means by which it is sustained. The reader who is familiar with the points controverted between Gall and Spurheim, will see how far these differences are solved by the view presented in my classification.

Those who commence to make Phrenological observations should first look at man as a whole, and notice the general form of the body, to see how the four temperaments exist in their general quantitative relations to each other. This practical instruction, after a certain number of observations are made, will enable the observer to see readily, in a given

Harmonious action of Brain, Stomach, Lungs and Liver important. Spiritual Faculties.

person, which functions of the system predominate, in respect to quantity and structural order, and what the relative influence of the brain, stomach, lungs, and liver, is in the organization; and thus to understand the physiological conditions to which the mental action is subordinated. Each of these functions, it should be remembered, is associated with an auxiliary apparatus, constituting a complete system which requires to be independently considered. Each of the four leading functions, the brain, stomach, lungs, and liver, may work either harmoniously or inharmoniously with the associated organs with which it is thus connected. To receive the highest development of human character, and to attain the end for which man was created, it is needful that each of the functions should be in just proportion, in size and activity, and have the same influence to the others. The brain should be properly related to the organs of special sense, so that sensuous life may be acute; the lungs should have the glottis, chest, and other parts of the respiratory apparatus with which they are associated, sufficiently developed, and structurally fitted to inhale the atmosphere necessary for the performance of their functions; the glands and ganglions should be in the right proportion, in size, to the stomach, and organs of nutrition, and properly expressed in form, giving symmetry to the whole structure of the body; and the liver—which has chiefly a chemical function, and is active in furnishing a supply of life-giving liquids with the proper gases, to the whole system, though it has also a characteristic organic force which the observer must not overlook—with its associated ducts, by which it performs its office of physiologic elimination, must be in due proportion to the other functions. It is to be remembered, however, that observation of size alone does not determine the *quantitative* action of an organ, but activity must also be taken into view, not only with the liver, but with all the other functions; and, finally, that it is the brilliancy of expression, or halo of the temperamental combination which manifests the existence and power of the Spiritual consciousness in its due predominance over the lower functions of the mind, which may express themselves merely in sensuous life.

The faculties composing the Spiritual group occupy the top of the head, and their place is designated on the bust by the words, "*Region of the Spiritual Faculties, Meditative and Intuitive.*" The Meditative Faculties, Steadfastness and Righteousness, are those which by inward conscious feeling of the right, make sentient the facts and truths already received into the Spiritual consciousness. They have a staid, natural restraining influence and character in the Spiritual group. If they predominate in this group, judgments will be given under restraint. If Godliness precedes their action, their influence will be prevenient, though only called out by necessity, when influenced by the power of the Holy Ghost.

These faculties give *Wisdom* as distinguished from *Knowledge*. A moral consciousness resident in these faculties is the mental condition of the highest powers and phenomena of human life in the natural state, these faculties of the brain forming the highest registering organism, including on the one hand, knowledge of Deity in outward manifestations by the faculty of Godliness, and on the other, the mental personality of the individual by the faculty of Self-Esteem and the Intellect.

The Intuitive Faculties are Brotherly-Kindness, Aptitude, Spiritual Insight, and Hope. These are the faculties which receive truths through spiritual impartation, by mediate transmission from outward objects and phenomenal aspects manifesting themselves to the subject. When influenced and directed by mere moral physical predominance of the organs of Godliness over the Meditative Faculties, Steadfastness and Righteousness, they give the capacity for Invention, Painting, Music, Oratory, and the broader artistic powers of Architecture and Sculpture,—all the poetic and literary gifts and graces which pervade civilization; in a word the mental form of genius.

This knowledge of the mind affords the essential means of solving those problems which have so long tasked the intuitive and meditative mind. The existence of evil; the sad and self-propagating disorders of human society; the inefficiency of external and social measures of reform; the historic tendency of National institutions to decline and fall; the fleeting nature of the civilizations of the past; and above all, the inefficiency of Christianity as now understood and received.

To know only these great characteristic facts concerning the race fills the mind with portents of its uncertain fate.

To understand these evils by a true analysis of the mind points out at once their causes, and the Divinely Given Power which is to be their cure.

Not the least consideration, prompting my labors in this direction, has been the hope, that my researches may stimulate others to investigate the great laws which regulate and control mental and physical life, and that their investigations may result in the rapid advancement and universal acceptation of sciences second to none in importance.

The diversity of mental manifestation leads us to conclude that every fibre entering into the composition of each distinctive convolution of the brain possesses a special operation. When the mind is more thoroughly illuminated, and its capacities sufficiently enlarged for more comprehensive and searching investigations, startling through well-ascertained discoveries of organs and functions—whose existence is now merely conjectural—will reward the searcher after truth. New and striking interest marks every progressive step. Especially to the Christian Philosopher is the field inviting, for, as these great laws are more thoroughly understood and practically applied, will the Spiritual not less than the temporal interests of mankind be promoted.

Respectfully yours,
JOHN HECKER.

APPENDIX.

THE WILL.

As there are voluntary actions which take place without Consciousness, there is also Sensation without Consciousness. If we use the word, therefore, in the broadest sense, to include the cause of all voluntary actions, whether conscious or unconscious, and without reference to the distinction between these two classes of volitions, we should characterize the Will as the function of the motor tract in the ganglia at the base of the brain. But in the more common signification, and the one more appropriate for our purpose, it connotes the Consciousness which takes place in the case of volitions which are preceded by and dependent on cerebral activity.

Metaphysicians, not considering the physiologic conditions of mental life, have not been able to agree upon the definition of the Will, some asserting that it was nothing more than desire, or as some have more nearly expressed it, the desire which is strongest at the time, thus describing only its mental character; while others have asserted that it is a distinct power or energy, not to be confounded with desire, thus describing only the physiologic element by contradistinguishing it from the mental.

Physiology enables us to define the Will more accurately as the cerebral determination of the activity of the motor tract. Any activity of the faculties controlling or attempting to control the bodily organism is an act of the Will.

It its mental aspect the Will has been well described as "the central point of the Consciousness." The mental quality of any volition is defined in the combination of faculties which unite in the act, and the peculiar character of those which are strongest in the combination.

When thus analyzed and organically explained, the Will gives the true constituent elements of the mental activity.

A.

APPENDIX.

The Will of any person, as distinguished from volitions, is the characteristic continuous phase of the successive conscious volitions. It is centralized in the habitual combination of the most active faculties.

Every act of conscious volition involves the expression of the three-fold phase of the mind, as I have before described it; force by the Propensities, knowledge by the Intellect, and either a dominant control, or a servient moral influence by the Spiritual nature.

Thus if the faculties at the base of the brain are large and active, the Will is characteristically sensuous. If the base of the brain is more contracted than the ranges of upper faculties the Will is characteristically qualified by a stronger infusion of moral and esthetic qualities. If the faculties of Self-Esteem and Steadfastness are large and active, the Will is characterized by a strong personality and inflexibility.

But since the activity of the faculties depends, not only on organic and temperamental conditions, but also on external influences, the mental quality of the volitions depends in part upon the circumstances in which the person is placed. At a time when the Will is centralized in Acquisitiveness and Secretiveness, for instance, the approach of a friend may modify the Will by the infusion of a larger influence of Approbativeness into the composite action of the faculties or may even supersede the predominance of the first named faculties, and cause the Will to be immediately centralized in Adhesiveness and other social feelings. At the next moment, the appearance of a person in distress may modify the Will by exciting Benevolence (Brotherly-Kindness) to predominant action, giving a very different volition, and next the approach of danger may intervene and the Will be instantly centralized in Cautiousness or Fear.

In this manner the mental phase of the Will varies constantly, the two conditions which determine the result being the organization of the individual and the surrounding external influences, either spiritual, or physical, or both.

The perverseness of the Will consists in its dependence on physical conditions instead of on Spiritual.

In the fallen state of man the external influences to which he yields himself are the physical and sensuous conditions. When the higher consciousness is subordinately exercised and the mind consequently is morally disposed, then the higher moral and esthetic qualities lead to abstraction in the special consciousness of the individual, only in connection with those mentally disposed like himself.

When man is born again, and the soul is awakened by the Holy Ghost, then only can the Will be centralized in God, this being the condition by which the truth is manifested in several persons, combined in the unity of the Spirit.

The perceptive observation of the organization of a person will inform us of the organic predispositions or habitudes of the mind in respect to the predominance of either group, and the phase of the temperament informs us what is the sensibility to external physical influences, and what modifications of mental action the bodily system tends to produce. But all these Perceptive facts of physical manifestation are seen to be superficial, when we regard the interior Spiritual conditions of the Truth. To discern human character, is a special gift of God, by the indwelling of the Holy Spirit, making the whole soul the instrument of His influence.

But God alone can foreknow human conduct because He alone, in whom all things move and have their being, foreknows the influential forces He has ordained.

From these facts respecting the physical conditions upon which the character of the Will depends, some persons ignoring the Spiritual power of God, have inferred a doctrine of necessity or fatalism, holding that man is purely the creature of circumstances, and hence denying moral responsibility.

And so indeed he would be if he were merely the creature that he is represented to be by those Physiologists and Phrenologists who have investigated the physical laws of his being, ignoring the Spiritual laws which I have discussed. When his life is that of the mere physical organization, he acknowledges merely a sentimental responsibility. It is only when the Holy Ghost is shed abroad in his heart that he is made consciously responsible to God. When his heart is exercised by the indwelling of the Holy Spirit manifested through his Spiritual group of faculties, he abnegates self, and looks to that Spirit for all he has to will and to do, in accordance with his Spiritual and moral attributes.

In truth, although a knowledge of the mind enables us to trace to their predetermined causes a much larger class of mental phenomena, including those of the Will, than self-consciousness subjectively exercised, would lead us to believe were so caused, yet the limited sphere in which the Will is free imposes a most cogent personal responsibility. The supposed tendency of scientific Physiology to supersede the doctrine of moral responsibility results from its not being believed that there is a Divine Power vouchsafed to the aid of Man's Spiritual nature. The organization and physical surroundings of every man are such, that at times his Propensities are quiescent or passive and the higher faculties for the time not wholly obscured by passions. This state is the condition of moral responsibility in its lowest degree. The full measure of moral responsibility is brought upon us by Christ and His sacrifice. Without dwelling on this point, it must suffice merely to say that this responsibility is single, individual, personal; and that it is direct and immediate to God himself without intervention of priest or church.

CONSCIOUSNESS.

The term "Sensation," in the physiological sense, which is the most appropriate signifies a *nervous* phenomenon. "Consciousness" denotes a *mental* phenomenon. To a considerable degree, Consciousness depends upon Sensation. But, as was explained in a previous letter, in which the physiological facts were traced in detail, Sensation may take place without inducing Consciousness.

Metaphysicians defining Consciousness have regarded it as that state in which any faculty is exercised; but they are not agreed as to what is a faculty, each presenting his theoretic analysis, without reference to the physical organization of the brain, by which alone we can individualize and define a faculty. Those affections which they name as faculties, Memory, Judgment, Hope, and the like, are, in truth, the activity of several faculties; so that it would be more correct to say that Consciousness is, commonly, if not always, the state in which two or more faculties are active.

There has been a difference of opinion, however, as to whether Consciousness is not itself a distinct faculty. In truth, when the mind reflects upon its own Consciousness, examining mental phenomena, the analytic faculties of the Intellect are exercised in the act, especially Individuality, or Eventuality and Time, or Causality and Comparison. It is from this introspective use of Intellectual faculties in examining the facts of Consciousness, that some persons have been led to regard Consciousness itself as an Intellectual faculty.

Phrenology solves for us the controversies of metaphysics as to the principles to be deduced from the facts of Consciousness, by showing that the character of Consciousness, in any case, depends on the character of the leading faculties which are concerned in the mental processes in question.

Hence, notwithstanding it should be admitted that the testimony of Consciousness in respect to the mind is uniformly true, yet the phenomena thus ascertained will necessarily differ in different minds, according to the faculties which are most active in the mental process, and in which the Consciousness of the individual may, therefore, be said to be centralized.

The Consciousness which is possessed by the mind in the natural, fallen state of man is that which is awakened by sensation, directly or indirectly. It is not in the power of Sensation, which is physical in its origin and nature, to awaken the whole mind to its full Consciousness. It can only awaken fully those faculties that are dependent on it—namely, the Propensities and the Intellect; and it influences the moral faculties, only indirectly by the sympathetic action of the Propensities and Intellect upon them, through their organic relations with each other, according to juxtaposition and predominance in size. Under this indirect influence, the moral faculties are not actively predominant; but may rather be said to lend their tone to the faculties of the other groups. They are only blind sentiments, and if Consciousness can be said to exist in them, it is not continuous, positive and dominant, but occasional, passive and servient.

The full power of Consciousness, that is the Consciousness of the whole mind, is only possessed when the soul is awakened by the Holy Spirit. It is the nature of the Spiritual Faculties, to respond, not to Sensation, but to Spiritual influence. God gives his Holy Spirit to quicken that which, by an apt figure, is said to be dead by nature. These faculties would always respond to this Divine influence were it not that, by indulgence of the Propensities, the mind is engrossed in the faculties which are connected with Sensation. The Consciousness which is awakened by Sensation is spoken of in Scripture as the natural heart,—the carnal mind,—the flesh. The Consciousness which is awakened by the Holy Spirit is designated the spiritual mind, the wisdom,—that is from above.

The higher Consciousness that is thus possessed is superior to that degree or phase of Consciousness which is awakened by Sensation. It includes the latter phase, and it can even control and suspend Sensation.

When our Consciousness is habitually centralized in these superior faculties, by dependence on the Spirit of God, then, consciously, "in Him we live and move and have our being."

In the early period of life, and to a great degree throughout life, except in the case of persons of exclusively intellectual pursuits, and the case of strong Christian experience, Consciousness is centralized in the Propensities.

Consciousness, when thus centralized in the passions, is active, forceful and commanding. One is not often uncertain as to the facts of Consciousness in the Propensities. These faculties are imperative; and if the mind examines its Consciousness in their exercise, the sense of liking or disliking is clear.

Consciousness in the Intellectual faculties is analytic and synthetic. In adult life of educated persons, the Intellect and the Propensities act so much together, that the Consciousness is commonly centralized in a combination of faculties, part of which are in each group. When the Consciousness is centralized exclusively in the higher Intellectual faculties, a comparative suspension of activity in the Propensities is necessary for the time being. In proportion as the leading Intellectual faculties acting, are higher in range, this power of abstraction increases; and hence the person may become for the time unmindful of bodily desires or inconveniences.

Consciousness in the moral faculties, in the natural, fallen state of man, is dormant and passive, not commanding. It is, as Dr. Spurzheim described these faculties, a blind sentiment. Even in the minds of persons trained under the influence of Christianity, and in whom the disposition of the other faculties is modified by the infusion of something of the tone of these, the only Consciousness which these faculties possess is still of this uncertain character, often persuasive, but not paramount nor controlling, and commonly so vague and indeterminate, that in self-examination the mind is at a loss to know whether it possesses any Spiritual Consciousness or not.

This is a very common phase even of Christian experience. A knowledge of the mind, will point the Christian to the physical and objective forms of the conditions which the Scriptures describe in spiritual language, as the basis of attaining this higher Consciousness. When the Spiritual faculties are fully awakened, and by watchfulness and prayer, the Propensities are held under due restraint, so that the predominance of the Spiritual faculties may be continuous, there is no longer vagueness and uncertainty in the Consciousness, but it is manifested, according to its own nature and by the witness of the Holy Spirit, as clearly and unmistakably as is Consciousness when centralized in the Propensities.

The complete Consciousness which is possessed by the soul, which has submitted itself to the grace of God, cannot be described and illustrated more aptly for my purpose than it is in the delineation given by Saint Paul in his Epistles, particularly the eighth chapter of Romans, and the second chapter of Galatians, and in external manifestation in his own conversion and his subsequent example.

We cannot, however, expect that Christians will attain the degree of the Spiritual Consciousness which he describes, while they neglect to unite in the spirit and order which Christ ordained, and combine their gifts as members of one body, in that union, as St. Paul has explained in detail.

Many theological teachers have confused themselves, and indirectly negatived the doctrine of the Holy Spirit, by confounding the processes of Intellection with the powers given by His influence. The Spiritual powers come from another source, and for another object, than those of the Intellect; and perceptive facts are not revealed by the Spirit, though He gives the power to know and use them.

Neither Intellectual methods of preaching, nor appeals to the social or personal affections, can directly influence the heart toward God. The important function of these methods, is incidental and preparatory, to modify and disarm the opposition of the Propensities and Intellect; and they can only prepare the way, in the mind of the hearer. The utterance of Spiritual truth, in the demonstration of the Spirit and with power, is given only when the Consciousness of the preacher is centralized in the Spiritual faculties, and his utterance is guided by the Holy Ghost.

In this analysis of the mind, and in examining the Scriptures with regard to it, we see too the error of those who have assumed that the Holy Spirit inspires the exercise of intellectual or passional powers. The wisdom which He gives, through the Spiritual faculties, is not the analytic or synthetic knowledge of the Intellect, though it uses this; nor are the affections which he gives the passional and sensuous disposition of the Propensities, though these also are enlisted in service.

The influence which He gives is primarily, and in its own nature, purely Spiritual, making Humility, Good Will, Judgment, and the other Intuitive and Meditative qualities the predominant leading phases of the character, by the conscious indwelling realization of the Spirit of God.

This is what the delineations of Scripture teach, describing the Spirit as enabling its subjects to worship and to love, to wait and to hope, to foreshadow the future, and interpret the obscure, to learn the truth and sell it not, to be pure, to discern and to judge; and by these powers fitting them to command and lead, not in pride but in humility.

In a state of barbarism, the Consciousness is ordinarily centralized in the sensuous faculties at the base of the brain, and chiefly in the Propensities; or, in other words, this is the part of the brain which is habitually exercised, and all the rest is more passive, being subject to the predominance of this lower range.

In civilization without a pure Christianity, the Consciousness ordinarily resides chiefly in the Propensities and Intellect, with such modifying influence as the blind sentiments of morality may give.

The forms of organization are more varying under the influence of civilization, than in barbarism, and, hence, the resulting manifestations of character are more diversified, but still the same general limit of the Consciousness is observable.

Consciousness in the higher Propensities and Intellectual faculties gives a worldly judgment and aptitude in the acquisition of wealth, the gratification of pride, the pursuits of ambition, the administration of jurisprudence, and in the exercise of the artistic and philosophic faculties, Ideality, Causality, and Comparison. This characterizes the leading elements in the present state of American Society.

SOCIAL ORGANIZATION.

In what has been said, in the foregoing letters, I have delineated the mental analysis of spiritual life, so far as it can be examined in the individual alone, and in the aspects of the relative proportional development and activity of the faculties of the individual; but, as we have seen in one of these letters, the activity of the faculties, especially of those in the Spiritual Group, is very largely sympathetic, dependent on the activity of other minds brought within mutual influence. The nature and extent of this susceptibility is one of the most obscure subjects in mental Philosophy, as at present usually taught; but the true science of the mind will disclose it to us by elucidating the conditions on which it depends.

The forms under which men are organized in society are, now, to a great extent, controlled by minor considerations resting in perceptive facts, and even what may be relatively called accidental circumstances.

In the first place, there is a certain relation between the mental organization of man as an individual, and the social organization in which he should stand. The essentially different order of the faculties in man and in woman, especially in the faculties of the Social and Animal Group, marks the fact that neither is complete, as it were, without the other, and that the first step in social organization is, as indicated by the first act of the Creator towards man, in this respect,—the giving him a help-meet. Husband and wife, united, each find, so far as individual and domestic life is concerned, the complement of their own characteristic nature supplied more or less appropriately in the other, and in the other alone.

The irregularities practised by men in this respect, since the creation of the race, mark the grossness of the passions, if they predominate and rule the whole mind.

In the family, the opportunities for the development of the children and the characteristic result of their development are very largely affected by the number of the children. If there be but one or even two children in the family there is great tendency to special and unequal development, especially if they are both of one sex. The faculties in the isolated child, which are originally predominant by hereditary causes, will be continually exercised beyond their due proportion, and the deficient faculties neglected, so that every thing tends to increase that which is large enough or too large, and diminish that which needs growth, and thus the angularity of the peculiar character is increased. A spoiled child is one in whom the passions are not regulated, within the child, by contact with others.

If, on the other hand, the family is very large, the necessary demands of the children divide the attention of the parents, and, if too large, will transcend the measure of attention which the mental organization of the parents is capable of maintaining ; and hence it will usually be found that, in a family of more than ten children, some will be neglected.

When men organize in society at large, the forms vary with the objects to be subserved. In general it may be said that the existing communities are mere aggregations of individuals, in forms which depend, in part, upon mental and temperamental affinities, but are rarely, if ever, based upon the true principles which a knowledge of mental organization establishes.

The basis of Political Governments is in the necessity of the communities in which they exist, which will continue as long as the Propensities of men predominate. When the Spiritual faculties predominate among men, then will be manifest the Kingdom of God.

THE CHURCH.

When our Lord established His Church on earth, He chose around Him twelve men as an organization for the accomplishment of His divine purpose, thereby intimating that this number should be the standard for all future associations whose object was Spiritual edification. In the union of such a number, special inequalities of individual development, whether excessive or defective, would be harmonized.

The seven faculties of the Spiritual group, with Godliness as the centre, and Brotherly-Kindness, Steadfastness and Righteousness in their proper order, with Hopefulness, Spiritual Insight and Aptitude, are self-adjusting and compensatory. The gifts of the various Faculties are distributed as the particular excess or deficiency requires, thereby producing harmonious and effective action of the whole.

The Propensities are essentially selfish in their nature, and the Intellect is centralized, and regardful of the highest individual advancement and personal interest only, hence, any inequalities in their development will exhibit harsh and discordant manifestations, which they possess no inherent power to pacify or abate. On the other hand, in the awakened Spiritual Faculties originate all the Christian graces which Christ Himself manifested. They are wholly unselfish and conciliatory, reflecting the divine essence, and manifesting, primarily, supreme love to God, and next, a warm sympathizing interest in the welfare of humanity.

But, in addition to the example set by our Lord, the facts of mental organization and experience would indicate that a number should be chosen, sufficiently large to be efficient and forcible, and the number of the Spiritual Faculties should be our guide in the formation of such a body. The discernment which the Holy Spirit gives for direction, and Spiritual sympathy being absolutely essential for the harmonious and effective accomplishment of the object, a number should be discerned and chosen, in which the unity, power and efficiency of the Church will be obtained.

Upon this point there are certain conditions of external facts in reference to the sympathetic activity of the Spiritual Faculties, which must be considered, as well as the subjective and individual conditions.

God's chosen people of old, the Jews, were raised up as a nation, consisting of twelve tribes, and governed in their tribes, families, and households, by Himself, through the law given by Him, and administered by men inspired and guided by the Holy Ghost. This theocracy had its systematic organization; and the law which God gave through Moses was adapted to such an administration.

The people, however, rejecting God, asked for a king; and although they were warned that a king would make the people tributary to himself, and would compel them to serve him, they persisted in their desire, and obtained a monarchical government.

The evils which Samuel predicted followed this departure from the theocratic government. From this time the prophecies of the coming of Christ, are to a great extent, characterized by presenting Him as their King, and by the promise of a restoration of the people, and that the kingdoms of the world should become His kingdom. When He came, He declared that He came not to destroy, but to fulfil the law. He required to be baptized, He took part in the service of the Synagogues. The cures which He performed, and which He declared depended upon the power of God manifested in Him, and upon the faith of the subject to receive the sympathetic influence, are consistent with what we know of the laws of mind and the power of the Spiritual faculties over the bodily conditions. After gathering about Him a number of disciples, He called them unto Him, and of them chose twelve, and ordained them that they should be with Him, and that He might send them forth to preach, and to have power to heal sicknesses, and to cast out devils. These twelve, selected from among all His followers, were His household. He taught that His kingdom was not of this world, but was spiritual. He forbade them to exercise authority, as the kings of the Gentiles, or to call any one master except Christ. Thus our Lord Himself re-inaugurated a free democratic theocracy.

This organization continued until after the resurrection and ascension of Christ. He taught them that He should found His Church upon the revelation of the Truth in the Heart of man, by God Himself; and to this end He promised them the Holy Ghost, who should guide them, His Church, into all Truth; and He declared, when He was about to leave them, that He sent them into the world as the Father had sent Him into it; that it was necessary that He should go away, thus removing the objective and physical presence of the personal manifestation of God, that they might receive the Holy Spirit in their hearts. And in reference to the miracles that He had done, He said: "He that believeth on Me, the works that I do shall he do also; and greater works than these shall he do; because I go unto My Father."

He left the Apostles to wait for the Holy Spirit. After His resurrection, and before He ascended into Heaven, He appeared repeatedly to the eleven when they were by themselves, and He breathed upon them that they might receive the Holy Ghost, and gave them power to remit sins, and finally commissioned them to " teach all nations and preach the Gospel to every creature, baptizing them in the name of the Father, and of the Son, and of the Holy Ghost, teaching them to observe all things whatsoever I have commanded you. * * * and these signs shall follow them that believe ; in my name shall they cast out devils ; they shall speak with new tongues; they shall take up serpents ; and if they drink any deadly thing it shall not hurt them ; they shall lay hands on the sick, and they shall recover; and lo, I am with you alway, even unto the end of the world."

But the Apostles, without authority, and without waiting for the gift of the Holy Spirit, resorted to lot to choose one to supply the place left vacant by Judas. They should have waited for the direction of the Holy Spirit, who Christ promised should come; and by His power overshadowing them, as it afterward did, they would have been guided ; but Peter, who was their chief, acted prematurely, being anxious no doubt to keep the number of the household of the Apostolic body full, through which Christ individually promised they should have greater power than He personally exerted when upon earth. Instead of thus resorting to chance, they ought, through Godliness and the other Spiritual faculties, and the special function of Spiritual Insight, guided by the Holy Spirit, to have discerned and tried the spirits of men whether they were of God.

Since that time, the organization, in the order in which Christ established it and left it, and directed its continuance through the predominance of the Spiritual gifts, has not been maintained ; and hence the powers which He promised that the Apostles as a household, and those which should believe through them, should possess, have ceased. The cessation of these gifts which are the fruits of the Spiritual power in their true order, is the objective proof that there has been a departure from that order, according to the test which Christ proclaimed—" by their fruits ye shall know them."

We see, in the present state of society, that the Church, instead of manifesting the essential characteristic of religion, that is a binding together, a tie of men to each other as well as to God ; and instead of possessing the powers of the spirit of unity, which He declared it should have, manifests separations and hostilities in the body of Christ, believers being divided into unsympathetic and rival orders, not having unity in the same spirit. Moreover in no one of them is there the organization or the spirit which Christ established and gave to His Church, and said His household should possess, as the characteristic of the body. He established a household of twelve, having all things in common, and ordained that the chiefest should be servant of all. The existing churches are organized without reference to these conditions of numbers, subsistence, and spiritual precedence. Those churches which dispute with each other the claim to have, by Apostolic succession, the right to teach, although practising ceremonial symbols of humility and mutual service, are actually organized on principles of lordship, authority, and preferment in temporal things. And those who disclaim the idea of any legitimate and orderly succession of the authority to teach, seem to maintain, to a greater or less degree, methods of organization which, in the same way, make the chiefest among them such as exercise authority over them, as Christ said His disciples should not.

APPENDIX.

If, therefore, it be asked, what is the proper organization of the Church, and how shall it be established, the answer which we draw from the organic nature of the mind and the teaching of the Scripture is, that in order to possess the powers which Christ authorized His Church to exercise, His disciples should be united in households of twelve, in the internal subjective unity of the spirit; and that, thus united, they should receive the external objective authority to teach, which should properly be derived through the most legitimate order of succession from the Apostles whom He commissioned.

To constitute the apostolic order as Christ ordained it, there must be—1. In the individual, the predominance of the Spiritual group of faculties, awakened by the Holy Ghost; 2. The organization of such individuals, chosen out from among men, to constitute the initiatory order of the true, practical, efficient, household of faith. (Organic predominance of the Spiritual faculties scientifically directs to this choice.) 3. The authority to teach, derived by such an household, by the most legitimate succession through some one of the Orthodox Catholic Churches. This authority is not an incidental matter of propriety, but is the condition upon which the fulness of God's power is promised.

Where, then, shall this authority be found? We must go back to the source of our title, if we would gain the authority and possess the power. This is an historical question. Which existing organization has, as an external objective fact, the clear antecedent? If it be the Greek Church, we must seek the authority there; if the Romish Church, we must seek it there; if the Church of England and its succession in the United States, we must seek it there; if any other order we must seek it among them. But wherever the clear chain of succession is, from thence the ordination of a household of twelve must be sought; and when the objective law is thus complied with, the manifestations of the spirit will be with power, as Christ declared they should be. His disciples will do greater works even than He did, and the promises of Scripture as to the evangelization of the world will rapidly progress towards their fulfilment.

However numerous these households might be, all their members would, necessarily, be of one mind, by the direction of the Holy Ghost, but with those diversities of gifts which arise from different mental organizations. It is on this individual diversity that depend the necessity and the beneficence of the peculiar unified organization which Christ founded. In such an organization the peculiar combination of faculties in the Spiritual group which might characterize each individual member would prove a special, spiritual gift, abounding in him for the edification, or strengthening of all the other members of the organization. The individual idiosyncrasies, too, which result from peculiar and special development in the Intellectual Group or the Social Propensities, and from peculiar temperamental conditions, would be harmonized. As society is now animated, the men who are deficient in the restraining faculties, are as a necessary consequence more or less dependent on the guidance and control of the selfish purposes of others, who are stronger in this respect. In the true order, those who now seem never to find their place, would come at once into perfect relations. Each member would suffer and would rejoice in the experience of his fellows; and the whole household would be, as the Apostle described it, one body of which Christ is the Head.

CONVERSION.

From what has been said above, it is apparent, *first*:—That the divinely intended character and life of man are attained in the individual, when he possesses, under suitable temperamental characteristics, the orderly development of the faculties of the brain, in their three classes or groups, among which, the Spiritual Group,—which is composed of the seven-fold spiritual gifts of Godliness, Brotherly-Kindness, Steadfastness, Righteousness, Hopefulness, Spiritual Insight, and Aptitude, under the influence of the Holy Spirit, must have predominance over the Propensities, Social and Animal, and over the Intellect; and, *second:*— That in social organization, the power and fulness of this life are to be attained, when individuals possessing this disposition are united, in Brotherly love, in the organization which Christ formed, and which is the condition upon which He promised the spiritual gifts to men. In this order of mental disposition, and to the extent of this social organization, men will be one with God, and one with each other.

We find, however, that in the actual state of man at large the predominance of the Propensities is the characteristic of his organization and life. This is shown, in History, in the life of Nations, in the present phenomena of Social and national life, in the objective, demonstrable facts of his physical organization, and in the inward and subjective consciousness of the individual, when that is regulated by Godliness, together with Steadfastness and Righteousness, the Meditative Faculties of the Spiritual Group.

A scientific analysis of mental operations, according to the objective, physical facts we have been considering, shows, that, in the state in which men naturally are, the Propensities rule them, and the Intellect serves the Propensities; and that the Moral or Spiritual Faculties are, as observers of the mind have described them, blind sentiments, unable to control. Instead of being ruled by Godliness, Brotherly-Kindness, Steadfastness, Righteousness, Hopefulness, Spiritual Insight, and Aptitude, and fulfilling the Divine law with all the strength in physical life that the due development of the Propensities gives, men are ruled by Alimentiveness, Amativeness, Destructiveness, Philoprogenitiveness, Inhabitiveness, Adhesiveness, Combativeness, Self-Esteem, Secretiveness, Approbativeness, Cautiousness and the Desire to live, with only such incidental modifying influence as the blind moral sentiments of Reverence, Benevolence, Firmness, Conscientiousness, Hope, Marvelousness, and Imitation may give.

This state of facts corresponds, as we have seen, to the delineation which the Scripture presents of man's nature. Revelation teaches us that man was created in the image of God, and at first existed in harmony with Him, enjoying the Divine presence; but that in the exercise of choice between good and evil, his Propensities, which are directly influenced by the external world, took the lead in the mind, subverting the predominance of the Spiritual Faculties, and thereby rejecting the influence of God, the Holy Spirit, who moves the soul to consciousness in those faculties. Thus was given to the whole mind and life, that gross, earthly, sensual, selfish character which is manifested by the predominance of activity in the Propensities.

Hence the race, instead of being able to live in peace and good will, under Spiritual guidance, have to be held in check as well as may be, by the Propensities of each other, regulated, to a greater or less degree in different communities, by those social laws which are but the expression of the Propensities of the community. Hitherto, these laws are only in part, and indirectly, infused with the spirit of Christianity, by God's good will.

The physical facts of man's organization indicate the connection between this perversion of nature, in which the Propensities are predominant, and the Death which is come upon the world.

APPENDIX.

Men in this natural state are characterized by selfishness and the passions. This selfish and passional nature is not always gross or offensive to men themselves. On the contrary, the Social Propensities manifest many qualities which tend to true happiness; and this is the function and effect of all of them, when they have their due relative order and proper direction. But when they are predominant, the man is under the immediate and controlling influence of things about him. Earthly desires, the sensuous knowledge sought to be procured through the Intellect when it is ruled by the Propensities, the evil influences which work in and with the Propensities,—these lead the Soul. This is "the carnal heart." This mental disposition, whatever may be its amenities and social graces " is enmity against God." The full and characteristic manifestation of the Intellect, when it is thus made the servant of the Propensities, is what the Apostle characterized as the wisdom which is earthly, sensual, [or natural] devilish." When the Spiritual Faculties become predominant, the heart is changed, for through them is given by the Holy Spirit, " the wisdom that is from above," which is " first pure, then peaceable, gentle, and easy to be entreated, full of mercy and good fruits, without partiality, and without hypocrisy."

To the phrenologic observer who understands the true nature of the Spiritual Faculties, this general, corrupt, depraved state of proclivity to evil is just as apparent to the eye, in the general predominance among men of the organs of the Propensities and the Sensuous faculties of the Intellect, and in the pantomimic and physiognomic indications of activity in the base of the brain, as the results of this proclivity in men's conduct are apparent to the moral sense.

As with the race, so with individuals, we are unable to rescue ourselves by anything which is in and of ourselves from this fallen state. We find no faculty or group of faculties in man, which, by their natural force, have the power to restore the true order of the mind. Therefore the need of the manifestation of the fulness of God by the Lord Jesus Christ, and the regeneration and renewing of the heart by the Holy Ghost.

The religious history of man shows that, down to the time of Christ, the comparatively few men who have received and yielded to the Spirit of God, have been ill-treated on that account, by their fellow creatures; and, especially, have those, through whom He has spoken to men, to recall them to Himself, been made sufferers and martyrs. The national history of the Jews presents this fact.

The only promise of relief that the world has seen is that which Christianity has brought; and under Christianity, although it is so imperfectly received, men have commenced a real progress towards permanent amelioration.

Christ came to manifest God to man, presenting in the perfect order, in human form, the Truth of God. And although it is not for us to assert any limits to God's grace, we know no other name under Heaven, given among men, whereby we may be saved. When men believe in the Lord Jesus Christ, who is God manifest in the flesh, they receive the Holy Spirit, who proceedeth from the Father and the Son.

In His conversation with Nicodemus, Christ described this change from the fleshly character given by the predominance of the Propensities, consequent upon the fall, to the spiritual character given by their due subordination to the Spiritual Faculties, under the influence of the Holy Ghost, as being born again: saying:—
" Except a man be born of water and of the Spirit, he cannot enter into the Kingdom of God. That which is born of the flesh is flesh, and that which is born of the Spirit is spirit. Marvel not that I said unto thee, Ye must be born again."

We are taught that our Heavenly Father is more ready to give the Holy Spirit to them that ask Him, than earthly parents are to provide for their children.

All that we are able to do is to negative ourselves, and present ourselves in a child-like humility as a living sacrifice, to God the Father Almighty, Maker of Heaven and earth, God's Spirit, the Holy Ghost, the Comforter, God the Son, the forerunner of the kingdom.

This humility is, as we have seen, the natural manifestation of the faculty of Reverence, or Godliness; and in order that it may characterize the mind, the demands of the Propensities must be checked or held under self-control.

This may be done by the will of the individual; and hence God commands all men every where to repent. Herein lies the freedom of the Will. And the responsibility and the duty is put upon every man, in this special act, to choose, by the resolution of the Will, between serving God and serving the devil. The mental processes which are involved in this change are as various as human organizations are various. Conscience, Fear, Love, Marvelousness, Gratitude, Hope, Reason, or even the Propensities when brought into great distress by unsatisfied desire, may call this faculty of Godliness into exercise, so that the man may centralize his will in humility, in place of pride and selfishness.

Those faculties which are most nearly contiguous to Godliness, will naturally be the ones to produce this effect; but when the Propensities are very strong, it is often the case that nothing but their own distress will lead the mind to this result. If the man is in immediate fear of the approach of death, and the Intellect can see no way of escape, the Desire of life, no longer able to be served by appealing to the Intellect, may appeal to the Spiritual Faculties. In the same way the suffering of other faculties in the Propensities, may bring the mind to state of willing itself to be humble, and thus prepare it to receive the Holy Spirit.

Bodily illness and pain, grief in what may be called the cluster of the household faculties, Philoprogenitiveness, Inhabitiveness, and Adhesiveness, caused by the loss of home, of the wife or husband, of children, and friends, the deprivations of the faculty of Alimentiveness by the want of food, and inability to obtain employment so as to procure food, may and should result in turning the mind to Spiritual life. It is thus that when the whole mental life has been centralized and engrossed in these lower parts of the brain, that the suffering incidental to their disappointment and deprivation is often the necessary condition of the awakening of the higher faculties.

In general, the faculties more remote in location from the Spiritual Group are less able to influence them than others. But whatever be the special process by which the result is attained, the result itself is essentially one and the same in kind; the submission of the whole heart to God, and the surrender of every one of the faculties in all the groups to His service, to love God with the whole heart, mind, soul, and strength.

It is very commonly the case, where the submission to the Holy Spirit is not complete, that the self-asserting power of the Propensities, or the too great predominance of the Intellect, prevents the full realization of this mental result. In this case the man finds within him opposing laws,—the purposes of his Spiritual Nature not having efficient control to guide the activity of the Propensities; and the Intellect analyzing this condition discloses to the inward consciousness the fact that the things which he would not, he does, and the things which he would, he does not.

APPENDIX.

This incomplete or imperfect change will be found to be more marked where the mental process involved in the change was characterized by the movement of the Propensities. If the person is brought to humility by over-wrought denunciations of his moral character, thus reaching the mind through Approbativeness and Self-Esteem; or if it is by fear only, presented by the approach of death or by physical images of future torment, thus reaching the mind through Cautiousness, Secretiveness and the Desire to live; or if it is chiefly by the social influence and solicitation of Christian friends, thus reaching the mind through Adhesiveness and Approbativeness; or in other similar ways; there is danger that the temporary exercise of the Spiritual Faculties under these appeals may be accepted, in the consciousness, as their actual awakening by the Holy Spirit and the consummation of the subjugation of the Propensities.

By whatever mental process we are brought to this state, we must, in order to give it the true and full efficacy, examine ourselves in the light of God's Spirit in us; for if the mind is truly and consciously in this humble state, He is revealed to the mind in the Faculty of Godliness, and Conscience bears witness of the fact to the soul. This it is to *know* God. Being thus revealed to us, both in ourselves and in the manifestation of Him in Jesus Christ as presented in the Holy Scriptures, He gives us a just consciousness of what we are in His sight. Hence comes that realizing sense of sinfulness and spiritual want which the heart feels when influenced by Godliness. This Consciousness is the inward manifestation of the change of heart.

Whatever other faculties may be exercised in this change, the mind can receive the Holy Spirit, only through the Spiritual Group,—Godliness leading the mind, and with it Brotherly-Kindness, giving love for mankind, and Steadfastness holding the mind in this state, Righteousness bearing witness to the Truth, and Hope giving aspiring anticipations for the soul's welfare, with the conscious indwelling Spiritual Insight of what is within, and Aptitude, by the Holy Ghost, subduing the whole man in harmony with God's Spirit, both in thought, will and deed. Thus Christ described it saying, "Whosoever shall not receive the kingdom of God as a little child, he shall not enter therein."

The language of Scripture, and the Religious history of man, will be found to correspond to and be explained by these physical objective facts of man's organization.

The Church is, as we have seen, the appropriate and appointed organism for the manifestation of the Holy Spirit in His fulness, and it is to be through the Church, when restored to the order in which Christ established it, that mankind, who now for the most part reject Him, will be constrained to receive the Holy Spirit.

THE STANDARD OF TRUTH.

From the nature of the Spiritual Faculties it will be seen that their right exercise is a matter of the utmost importance to man. This is the "one thing needful."

The great practical question is. "What is Truth?"

The discordance which exists among men in this respect is due to the causes that lie in the inherited predisposition to activity in the Propensities, which is the origin of the sinful disposition of man, and the natural blindness of the Spiritual faculties, causing the failure of man to receive the real standard of Truth. Men by nature live in the Propensities, and these seek always their own gratification.

These faculties know nothing higher than their objects of desire, which all centre in or relate to self; and they not only are not ruled by the Truth, but their own prior, predetermined activity excludes it, when it conflicts with the immediate gratification which they seek.

The Intellect may logically define and so discern Truth in outward, objective forms, but it is ruled by the Propensities, unless they have been subordinated to the Holy Spirit through the Spiritual Faculties. By the Holy Spirit man may know the Truth and be redeemed. The Scriptures state this Spirit to be the power by which Christ was conceived. The power He manifested unto man He always declared to be by the instrumentality of the Holy Ghost.

Even when the Spiritual Faculties have gained predominance, so that the sinful disposition has been changed and the Propensities are under control, and men seem sincerely to seek the Truth, discordance remains among them; and this is because they do not admit the Holy Spirit to be the standard of Truth, which alone can harmonize the mental diversities among themselves; and this Spirit of Truth can not be received in His fulness, except when men are united with each other in Him, in the order which Christ established. This order was lost when the Apostles, led by Peter, resorted to lot to choose one to be of their number, instead of relying upon the wisdom and insight to be given by the Holy Ghost; and the disorder entailed upon the Church by this act still prevails. God Himself, the Holy Spirit, is present with man ready to guide him into all Truth. But even Christian men, being deceived from the want of the order above described, and thus being left to their own opinion, err so far, as to assert other guides and standards.

Ecclesiastical authority, which some assert to be the true standard of opinion is divided against itself; and there needs a standard to ascertain which power, if any, is genuine and which specious.

The doctrine of the *Right of private judgment* affords no standard of Truth. It is only in an indirect sense that there is such a right. There is no right to hold error; but only a right not to be molested by others for holding error. It needs the true science of the mind to rescue man from the fallacies and self-deception of individual opinion.

The Scriptures, which are often asserted to be the perfect standard of Truth, do not ascribe this character to themselves, but to the Holy Ghost. The Scriptures are a record; and although made by men of like passions with ourselves, they were written by "holy men who spake as they were moved by the Holy Ghost," and they are profitable for instruction, for reproof, and for doctrine, and we are to search them if we would have eternal life, for they are they which testify of Jesus Christ. This record presents the history of the influence of the Holy Spirit upon men in times past, and it is given to us that we may be led to receive the influence of the same Spirit. A record of the past cannot in its nature be adequate in all its conditions as a guide for the present and future. It cannot be everywhere and at all times accessible. The gospels were kept away from men for many hundred years, and still are, in many countries; and Protestantism has not yet wholly succeeded in rescuing the Bible for men. Nor is a record intelligible to every mind; and the interpretations put upon it among men, unless there be some other standard, will differ.

Almighty God the Father, and Jesus Christ His only Son, and the Holy Spirit proceeding from the Father and the Son, are the conditions of the Truth. Hence the Scripture and Christ Himself teach us to regard the Holy Spirit, as the immediate interpreter and standard of the Truth in the Spiritual Faculties properly organized to receive Him. Christ promised to his disciples the gift of the Comforter " that He may abide with you forever,(even the Spirit of Truth); whom the world cannot receive, because it seeth Him not, neither knoweth Him; but ye know Him; for He dwelleth with you and shall be in you." * * * " He shall teach you all things. He is "the Spirit of Truth which proceedeth from the Father." " He will reprove the world of sin, and of righteousness and of judgment." " When He, the Spirit of Truth, is come, He will guide you into all Truth."

And again, the Scriptures say of those who live in God as His children, " Ye have an unction from the Holy One, and ye know all things," " The anointing which ye have received of Him abideth in you, and ye need not that any man teach you: but as the same anointing teacheth you of all things, and is Truth, and is no lie, and even as it hath taught you, ye shall abide in Him."

And again—" Now therefore ye are no more strangers and foreigners, but fellow-citizens with the saints and of the household of God. And are built upon the foundation of the apostles and prophets, Jesus Christ Himself being the chief corner-*stone*; in Whom all the building fitly framed together groweth unto an holy temple in the Lord; in Whom ye also are building together for an habitation of God through the Spirit."

Christ, when upon earth, was the only *living standard* of Truth that the world has ever known. The *Holy Scriptures* are the only *outward or objective standard*; and they maintain their place and authority through the ages, illustrating that those things are most permanent which are the most immediate work of God:

The real *subjective standard* is the HOLY SPIRIT manifested in man. The Spiritual Faculties, existing in their true order of development, and awakened by the Holy Spirit, receive and manifest the Truth to the whole mind. They acknowledge God, as all in all, and give Humility to man. They manifest Love to man giving Peace on earth. They maintain Faithfulness, Uprightness, and Purity of life. They alone give the Hope which maketh not ashamed. They give Judgment, (that which arises from the Meditative Faculties) that wisdom which God promises to those who have Faith in Him and ask Him, and which is superior to the highest analytic power of the Intellect. They subordinate the Desires of the flesh, holding them in check, and directing them in rightful and healthful activity; and they illumine Knowledge, bringing every Intellectual Faculty into harmony with the Truth.

When men perceive the facts of mental organization, they will understand the ground and the causes of differences of opinion; and when they accept the Holy Spirit as the Spirit of Truth proceeding from the Father and the Son, and the standard for them, they will subordinate themselves to Him and so will unite in the Truth as He manifests it.

www.ingramcontent.com/pod-product-compliance
Lightning Source LLC
Chambersburg PA
CBHW022009220426
43663CB00007B/1022